CP001841
CORNWALL COLLEGE

# RAPID TOOLING

# RAPID TOOLING

## Technologies and Industrial Applications

edited by

**Peter D. Hilton**

***Technology Strategies Group***
***Concord, Massachusetts***

**Paul F. Jacobs**

***Laser Fare—Advanced Technology Group***
***Warwick, Rhode Island***

MARCEL DEKKER, INC. NEW YORK • BASEL

**ISBN: 0-8247-8788-9**

This book is printed on acid-free paper.

**Headquarters**
Marcel Dekker, Inc.
270 Madison Avenue, New York, NY 10016
tel: 212-696-9000; fax: 212-685-4540

**Eastern Hemisphere Distribution**
Marcel Dekker AG
Hutgasse 4, Postfach 812, CH-4001 Basel, Switzerland
tel: 41-61-261-8482; fax: 41-61-261-8896

**World Wide Web**
http://www.dekker.com

The publisher offers discounts on this book when ordered in bulk quantities. For more information, write to Special Sales/Professional Marketing at the headquarters address above.



Current printing (last digit):
10 9 8 7 6 5 4 3 2 1

**PRINTED IN THE UNITED STATES OF AMERICA**

# Preface

*Rapid Tooling: Technologies and Industrial Applications* describes the current, albeit quickly evolving, state of rapid manufacturing (RM) and rapid tooling (RT), and identifies the basic aspects of each commercially available RP&M system. The primary goal of this book is to provide useful information to individuals and organizations considering the use of rapid tooling technologies in product development. It discusses the benefits of using rapid prototyping and manufacturing (RP&M) technologies in the development process, and identifies complementary technologies—such as computer-aided design (CAD), computer-aided engineering (CAE), and computer-aided manufacturing (CAM)—that need to be applied in conjunction with RP&M to achieve maximum benefits.

The book is written for people who need to determine whether, or when, to introduce RP&M into their organization. Engineering managers responsible for product development or manufacturing processes should find this book extremely valuable in providing a background for the use of RP&M within their organization. R&D managers with product development responsibility will find information regarding advanced techniques that their departments will want to assess, and quite possibly introduce and support. Mechanical engineers, material scientists, and manufacturing/industrial engineers who may be called on to use RP&M technologies should find specific information within this book that is directly relevant to their work. Finally, RP&M technology and business participants will want to read this book to learn more about the state of the technology, some of its unique applications, and the likely direction of its future development.

The RP&M industry has shown various signs of maturation. There has already been industry consolidation. Several firms have developed essentially stable market positions, one system supplier has failed, and several others are losing ground. Rapid prototyping services are available from over 350 service bureaus worldwide. The competition for business among these firms has driven prices downward, and reduced profit margins. Somewhat ironically, the low prices of RP&M parts that have adversely impacted some service bureaus have caused other organizations to use service bureaus rather than purchasing, installing, and operating their own RP&M equipment internally. This, in turn, has adversely affected the sale of equipment from the RP&M system manufacturers. As a consequence, revenue growth has slowed, or even reversed, and losses have often replaced profits in the quarterly reports of these firms.

Notwithstanding all these difficulties, the picture is hardly as bleak as one might initially surmise. At its core, RP&M is really about catching errors early in the design process, designing better products, reducing product cost, and getting products to market faster. All four of these benefits are surely coveted by nearly all industries. We believe that the first order of business for the RP&M industry is *education*—which is one of the primary reasons for writing this book. As more companies learn about the time and cost savings that are possible with RP&M, business opportunities for our industry will expand. Although awareness of RP&M is certainly much greater than it was a decade ago, the percentage of those companies that manufacture a physical product while utilizing RP&M remains pitifully small. To get a sense of this, the next time you attend a dinner party or a baseball game and the person next to you is an engineer, scientist, or business manager, ask that person if they have ever heard of RP&M.

The old saw ''nothing succeeds like success'' is truly relevant. Those organizations that have experienced significant time or cost savings, or improved product quality through RP&M, become ''true believers.'' These firms continue to use the process over and over again. What does it take to convince someone who has not achieved these benefits that they are real? How does one show someone that these benefits can be applied to his or her specific application? Is the lack of adoption related to fear of failure? Are these people afraid that if they recommend the use of RP&M during the kickoff meeting for their next product development, others will look at them as if they were from Mars? Perhaps if they read the story of ''Project Widget'' in Chapter 3 they will realize what they might be missing if they do not utilize RP&M.

We believe that real success stories documenting genuine benefits are key to expanding the adoption of RP&M. In fact, we believe this so thoroughly that we have included such case histories in this book from organizations willing to share the details. The real growth potential for the RP&M industry lies not with 1% of companies currently using the technology but with the 99% who have yet to do so!

Rapid manufacturing, and specifically rapid tooling technologies, are earlier in their development than rapid prototyping (RP) technologies, and indeed are often extensions of RP. New technology advances, such as laser engineered net shaping (LENS), are continuing at a rapid rate. Certainly, no single technology can do everything. All the current RT processes have some limitations that compromise their broad adoption. We believe that further research and development in these areas can and will enable one or more of these methods to gain a foothold.

The current tool and die industry is estimated to involve annual revenues of roughly $10 billion. The plastic injection molding market is estimated at about $20 billion per year. Thus, the opportunity for growth into these areas is very significant. We firmly believe that the time, cost, and part quality benefits associated with the methods described in this book are substantial, and that possibly within five years, and almost certainly within 10 years of this publication, alternative tooling techniques will account for revenues exceeding $1 billion per year. Interestingly, this would exceed the entire RP&M industry revenue for 1999.

Chapter 1, by Peter Hilton, describes the current state of rapid manufacturing, including a brief summary of the major commercial approaches. Chapter 2, by Georges Salloum of the National Research Council of Canada, addresses the relatively broad topic of computer-based tools used in product development. This chapter focuses on the use of CAD and CAE simulation to evaluate product functionality as well as the processes used to manufacture the parts.

Chapter 3, by Paul Jacobs, presents a fictional product development story intended to illuminate the problems, challenges, and opportunities that currently exist in order to significantly reduce the product development cycle. The story, "Project Widget," highlights many of the various issues associated with product development including concept definition, task management, the role of suppliers, initial prototyping, tooling requirements, and manufacturing process development. The chapter provides a background for the use of RP&M throughout the book.

Chapters 4 through 8 address alternatives to conventional machined prototype or production molds, including various methods to accomplish soft tooling, bridge tooling, cast tooling, and production tooling. Chapters 9 through 11 focus on specific applications of RP&M that are currently being employed in the automotive, medical device, and investment casting industries. The final chapter, by Peter Hilton, provides a perspective on the future of RP&M, addressing its likely market penetration and technology growth.

Peter Hilton would like to dedicate his efforts in the publication of this book to his wife, Joannie Hilton. Paul Jacobs would like to dedicate his efforts in the publication of this book to his parents, Margaret Veronica Jacobs (1910–1999) and Bertram Lawrence Jacobs (1899–1975).

## ACKNOWLEDGMENTS

Paul Jacobs would like to acknowledge the following individuals whose teaching, counsel, wisdom, effort, support, and vision have helped shape a career:

- Gardner Ketchum, Raymond Eisenstadt, and Carl Niemeyer at Union College
- Jerry Grey, Martin Summerfield, and Lyman Spitzer at Princeton University
- Gordon Cann, Rolf Buhler, Ken Gustafson, and Bill Hug at Xerox Corporation
- Chuck Hull, Hop Nguyen, Rich Leyden, and Jouni Partanen at 3D Systems
- Terry Feeley, Kip Brockmyre, and Tom McDonald at Laser Fare.

*Peter D. Hilton*
*Paul F. Jacobs*

# Contents

# Contributors

**Anthony T. Anderson** Ford Motor Company, Redford, Michigan

**Daniel L. Anderson** DePuy Orthopaedics, Warsaw, Indiana

**Larry André, Sr.** Solidiform, Inc., Forth Worth, Texas

**Debbie Davy** Mirotech, Inc., Toronto, Ontario, Canada

**Peter D. Hilton** Technology Strategies Group, Concord, Massachusetts

**Paul F. Jacobs** Laser Fare—Advanced Technology Group, Warwick, Rhode Island

**Hugo Lorrain** Howmet Aluminum, Laval, Quebec, Canada

**Thomas R. Richards** American Industrial Casting, Inc., East Greenwich, Rhode Island

**Georges Salloum** Integrated Manufacturing Technologies Institute, National Research Council of Canada, London, Ontario, Canada

**Sean Wise** CEMCOM Corporation, Baltimore, Maryland

# RAPID TOOLING

# 1
# Introduction

**Peter D. Hilton**
*Technology Strategies Group*
*Concord, Massachusetts*

This book focuses on the manufacturing portion of the broader *rapid prototyping and manufacturing* (RP&M) field. Our interest is in the rapid production development of relatively low-volume functional parts: parts made out of the production materials and produced by the production processes. Examples include investment-cast, nickel alloy aerospace engine components and injection-molded polymeric parts (e.g., electronic enclosures). Developing the ability to produce these parts requires developing forming molds or tools for the parts. Traditionally, the development of such molds or tools is by machining and heat treating; it requires substantial calendar time and has significant associated costs. Further, changes to the molds and tools also require significant time and costs. Therefore, it is of interest during product development to be able to quickly produce some first ''real'' parts and to be able to modify the subsequent parts rapidly based on findings associated with these first parts. We call the ability to rapidly develop molds or tools for moderate volume parts or products *rapid manufacturing*.

## I. CONTEXT FOR RAPID MANUFACTURING

Rapid manufacturing (i.e., the rapid production of molds or tools) can be accomplished throught the use of some rapid prototyping processes followed by some subsequent processes. For example, an RP model of the part sought can be produced and subsequently used as a sacrificial pattern to investment cast the part. Alternatively, a mold can be designed and the patterns for making the mold can be produced in plastic or wax using an RP technology. These RP pieces can be used sacrificially in the investment-casting process to form mold inserts in metal.

The various rapid manufacturing processes (to be discussed in this book) compete against computer numerically controlled (CNC) machining. CNC is the more mature technology that is threatened by the newer RM technologies. As is not untypical in these situations, advances are being made in CNC and related technologies in response to the threat.

It is interesting to postulate how long it will take for RM processes to replace traditional toolmaking processes. Material presented in this book will show that RM processes are still under development. It is premature to select the winning technologies, although some of the losers are already becoming obvious. The most recent work indicates that we are able to produce molds for high-volume production using some of the alternative technologies. The technologies require further field verification to develop the needed confidence in their long-term performance. However, they are able to contribute to addressing the critical competitive factors of time and quality through reducing product-development time, improving productivity, and enabling product dimensional control quality.

The benefits of a new technology are always weighted against the risks. Leading users are those who are willing to implement the technologies early, assuming risks in the hope of achieving competitive advantage. These firms typically have strong technology competencies and are able to survive start-up glitches.

The rate of technology acceptance varies enormously by technology category and application industry. New electronics technologies that provide competitive benefit, particularly software, are generally implemented very rapidly. Materials-related technologies, particularly for transportation applications, typically require decades to achieve substantial market penetration. Rapid manufacturing technologies have attributes of both software and materials processing technologies. The authors anticipate that the market penetration of RM beyond the lead users will be quite slow though steady. Our projection is based on our sense that the mold-making industry is slow to embrace change and that they will need to be pushed by their OEM (original equipment manufacturers) customers to implement new technologies.

## II. THE PRODUCT-DEVELOPMENT PROCESS: REDUCING TIME AND COST WHILE IMPROVING PRODUCT FUNCTIONALITY AND PRODUCIBILITY

The strong interest in RP&M stems from a more broad directional change in industry toward more rapid product development. There are numerous reasons for wanting to develop products more rapidly and a great deal of pressure to do so. Examples where product-development cycle time pressures are well known include automotive, where the time to develop of a new car is being reduced from approximately 60 months 10 years ago to 18 months today. The shorter the development time, the more effectively the developer can respond to current or recent consumer trends (e.g., for sports utility vehicles). In electronics, product cycle times are being driven down to less than 1 year, requiring very rapid and cost-effective development. Toys need to be developed during the first quarter of the year for full-volume production in the third quarter to enable sale during the fourth quarter. Many, if not most, product areas are now under pressure for rapid product development.

It is not enough to develop products rapidly. The products clearly need to be attractive in terms of the market drivers and the processes for manufacturing them need to be both robust and cost-effective.

## III. ADVANCED TECHNOLOGIES SUPPORT RAPID PRODUCT DEVELOPMENT

The sequence of the product-development phases and the overlapping functional roles are illustrated in Fig. 1. Rapid and effective product development requires a number of capabilities, including an effective rapid-product-development process, strong competencies and resources, and supportive management. There are several areas of advanced technology specifically developed to aid rapid product development; these include the various computer-driven tools, computer-aided design (CAD), computer-aided engineering (CAE), computer-aided manufacturing (CAM), RP&M, and virtual prototyping of both product functionality and the processes for manufacturing and assembling the product. These tools aid in product design, analysis, prototyping, simulating, and manufacturing process development.

Integrated engineering software and electronic communication with internal and external participants reduces the time and cost of product develop-

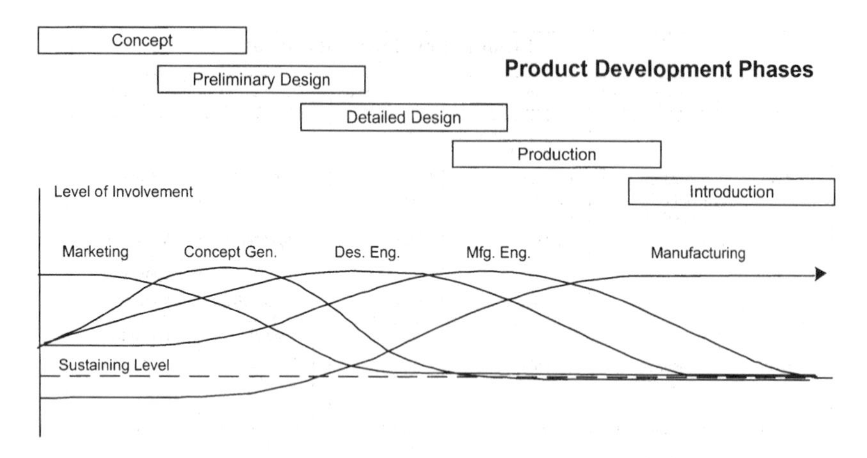


**Figure 1** Product-development strategy. Concurrent product development links the product-development team with the plant and facilitates effective product transition into manufacturing.

ment while minimizing error introduction. Figure 2 lists the computer tools used to support product development and their roles during the various phases of the development process. The product and its components are designed on a CAD system. The CAD models are transferred to a CAE environment for analyses of product functional performance and of the manufacturing processes for producing the product. The CAD information is also transferred to those responsible for manufacturing process development and they use it to design tooling and to create the CAM files for machining operations. The CAD file is transferred once more to those who will produce prototypes and patterns using RP. The participants in these various processes may be internal to the company developing the product or they may be external suppliers of tools, RP services, or analysis services. Thus, the integration of these engineering software systems to enable direct communication between them as well as the electronic communication network among the product-development participants provides important leverage in the product development process.

Virtual prototyping is the natural extension of CAE (engineering analysis). It simulates the product functionality and the processes for producing it prior to development of physical prototypes. Virtual prototyping enables the design team to perform at least one design iteration without producing hardware—thereby saving time and cost. Virtual prototyping tools also guide in optimization of the product and the manufacturing process.

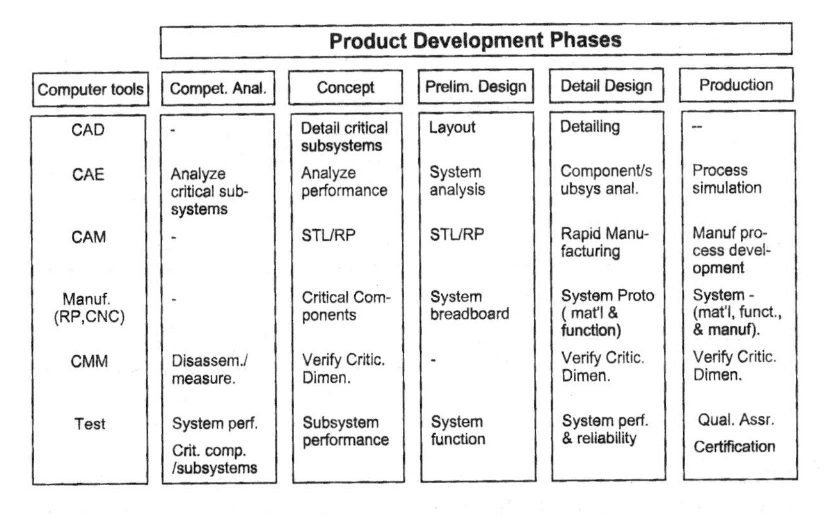


**Figure 2** Integration of computer tools for improved product development.

Physical prototyping enables physical contact with the proposed product by various interested parties, including the design team, the manufacturing department (concerned with how to produce the product), the marketing department (concerned with the products appeal), and potential customers (concerned with whether the product can perform the functions they have in mind). The physical prototype may also serve as a pattern for a forming process by which replicate parts are produced.

Rapid manufacturing actually refers to two functions—the rapid development of "tooling" for the conventional manufacturing process (e.g., molds for injection molding) and rapid-manufacturing cycle times (e.g., conformal cooling of molds to reduce the injection-molding cycle time). Both functions provide competitive benefits. Some RP&M processes contribute to both reducing the development time and the cycle time.

## IV. A BRIEF REVIEW OF RAPID-PROTOTYPING TECHNOLOGIES

The history and present state of rapid prototyping is reviewed next. The reader is referred to several textbooks (1,2) for more information on the subject of rapid prototyping.

Rapid-prototyping technologies have been commercialized over the last 10 years following inventions by Charles Hull, founder of 3D Systems, and others. The original concept for the application of RP was to quickly produce geometric prototypes from CAD files early in the product-development cycle. Charles Hull developed and commercialized stereolithography (SL), a process by which a photosensitive polymer is cured selectively by an ultraviolet laser beam. The laser beam is moved along a path defined by a computer model to create a two-dimensional pattern of cured resin; the resin bath is lowered, causing a layer of liquid resin to cover the cured layer; and the process is repeated, resulting in a three-dimensional (3D) layered object.

Other processes have also been developed which enable a wider range of materials to be used. A process, known as selective laser sintering (SLS), was developed at University of Texas, Austin and licensed to DTM Corporation. In SLS, a laser moves over a layer of polymeric powder in a pattern controlled by a computer, causing local sintering (heating and adhesion) of the powder. As with stereolithography, the platform is lowered, another layer of powder is distributed over the previous layer, and the laser beam is scanned over that layer to form the next layer of the prototype part. SLS is applicable to thermoplastic resins, waxes, and thermoplastic-coated metal powders.

Professor Sachs at MIT led the development of a rapid-prototyping approach based on ink-jet printing technology. The process is known as 3D Printing. Printing heads apply drops of adhesive in a pattern on to a bed containing a layer of powder; the bed is lowered; a new layer of powder is distributed; and the printing process continues. 3D Printing can be applied to resin powders, ceramic powders, and metal powders. In the case of metal powders, subsequent sintering and infiltration are needed to form a fully dense metal object.

3D Printing technology has been licensed to several firms for distinct application types. Soligen uses 3D Printing to form ceramic shells for investment casting; Extrude Hone uses the technology to directly form metal tools (sintering and infiltration are still required.); and Z-Corp produces relatively crude prototypes very rapidly. The Z-Corp machine uses multiple jets to increase fabrication speed.

Still other rapid-prototyping technologies are based on distributing a thermoplastic resin in a heated, viscous state. Stratasys applies a ribbon of resin through a moving extrusion head under computer control to create two-dimensional (2D) layers, each on top of the previous layer, to form a 3D model. Sanders (Sanders Modeler) uses ink-jet technology to apply the thermoplastic resin in a liquid state. Helisys uses a $CO_2$ laser to cut paper sheets that are adhesively bonded to form a layered structure. Cubital uses an ultra-violet-sensitive polymer and a photo-transfer-based approach to create each complete layer concurrently. Ballastic particle modeling technology directs polymeric particles at high velocity onto the target, where they adhere to form a 2D pattern and repeats the process to cause the formation of a three dimensional layered part.

The major commercial technologies are listed in Table 1, followed by their associated equipment producers, material capabilities, and applications. Sandia National Laboratories is developing a process to directly produce solid metal parts. The process, known as Laser Engineered Net Shaping (LENS), involves the use of a high-power YAG laser that locally melts the metal substrate while metal powder is directed into the resulting melt pool. As with the previous processes, LENS forms the three-dimensional object by creating a series of layers.

Each of these rapid-prototyping techniques uses a similar approach to transfer the part geometry. A solid model of the part is developed on a CAD system and subsequently sliced by parallel planes to create a series of equal-thickness layers. Each of the rapid-prototyping systems recreates these layers in physical material and builds up the layers to form the part. The distinctions among the systems are the process for putting down material and the materials applied.

Recent technology development in rapid prototyping includes both the refinement of existing processes and the development of new processes such as LENS. Both 3D Systems and DTM have been adding new materials for their RP processes. 3D Systems with the help of the resin producer, Ciba Geigy, has been refining resin systems to improve toughness and dimensional control. They have also added control algorithms (build patterns) to enable the creation of a ''honeycomb'' structure with continuous surfacing. This process, known as QuickCast, provides advantage when the RP part is used as a casting pattern. The developments in resin systems in combination with build pattern refinements have enabled 3D Systems to substantially improve the accuracy capability of stereolithography. In addition, 3D Systems has developed new models of their rapid-prototyping equipment which have increased forming speed and enable larger RP parts. DTM has added two new material types to its RP capabilities; one is a proprietary resin system called TrueForm. Its use enables more accurate parts and more effective casting patterns. The second new powder is metal encapsulated in a polymeric coating and the associated RP process has been named RapidTool. RapidTool, as the name suggests, represents a process of directly forming metal parts that can be used for tooling applications.

**Table 1** Listing of RP Technologies, Companies, Materials, and Applications

| Technology | Equipment manufacturers | Materials capability | Applications |
| --- | --- | --- | --- |
| Stereolithography | 3D Systems | Epoxy resins | Prototyping, casting patterns, soft tooling |
| Selective laser sintering | DTM | Thermoplastics, waxes, metal powders | Prototyping, casting patterns, metal preforms (to be sintered and infiltrated) |
| 3D printing (binder printed onto powder layer) | MIT, Soligen, Extrude Hone, Z Corp | Metals, ceramics, other powders | Prototypes, casting shells, tooling |
| Laminated object manufacturing | Helysis | Paper | Prototypes, casting models |
| Fused deposition modeling | Stratasys, Sanders | Thermoplastics, waxes | Prototypes, casting patterns |
| Solid ground curing | Cubital | UV curable resins | Prototyping |

Recently 3D Systems commercialized a second type of rapid-prototyping technology that enables it to produce fast prototypes at lower capital costs. This prototyping machine called Actua and even more recently ThermoJet uses multiple heads to extrude thermoplastic resins and form prototypes rapidly. The technology competes with the Z-Corp technology. It does not significantly overlap stereolithography and therefore represents an extension of the rapid-prototyping functionality offered by 3D Systems.

Rapid prototyping has appeal and is widely used in industry. However, the need for physical models may have peaked at some of the technology application leaders. Computer simulations and virtual prototypes are replacing some early physical prototypes in the product-development cycle. Whereas performance simulation and virtual prototyping may become a threat to the prototyping business, rapid-prototyping technologies are now finding growing applications beyond prototyping.

## V. VIRTUAL PROTOTYPING OF PRODUCT FUNCTIONS AND MANUFACTURING PROCESSES

The purpose of prototyping a product during development is to give the various interested parties (including engineering, sales and marketing, manufacturing, parts suppliers, and subcontractors) a better sense for the product. The prototype can serve to demonstrate functional attributes of the product, to exhibit its appearance, or point out manufacturing issues or requirements. Advances in computer simulation enable much of these purposes to be addressed through virtual prototyping. The advantages of virtual prototyping are illustrated in Fig. 3, which shows the physical prototyping steps that can be replaced by computer simulations. Solid modeling in combination with appearance-enhancing software can create attractive images of the product. Various analysis and simulation packages enable assessment of product functionality; for example, kinematic modeling enables motion simulation, CFD (computational fluid dynamics) can replicate a wind tunnel and assess fluid flow, FEA (finite element analysis) can be used to determine load-carrying capacity and to predict temperature distributions. Other analysis tools simulate various manufacturing and assembly processes; for example, software packages are available to model most of the usual materials-forming processes such as injection molding, investment casting, closed die forging, and so forth. Other packages simulate the various assembly operations, providing insight on setting

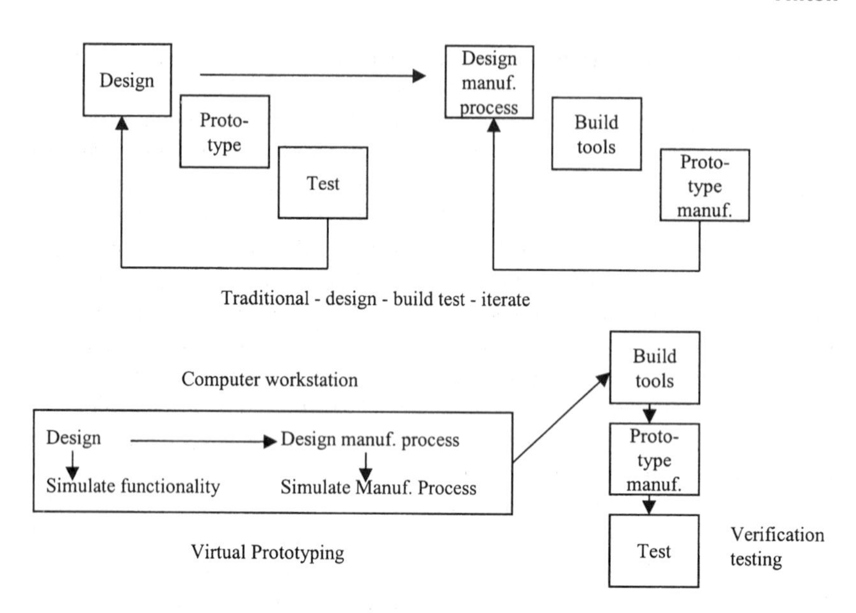


**Figure 3** Comparison of traditional (new product development) with testing to new product development with virtual prototyping.

up a manufacturing line. Georges Salloum discusses computer simulation and virtual prototyping in Chapter 2.

## VI. OVERVIEW OF RAPID MANUFACTURING

Rapid-prototyping technologies are being used to create patterns for casting processes—for urethane casting and for investment casting of metals. In the case of urethane casting, the RP piece is the pattern for producing a silicone rubber mold that, in turn, is used to cast a number of urethane parts (typically 1-to-50). Urethane casting is an effective process when one needs to create multiple prototypes for evaluation purposes. For investment casting, the RP piece is used in a sacrificial manner in place of the traditional wax pattern. It is coated with a ceramic slurry that forms a shell. The RP piece is melted or burned out. Molten metal is poured into the shell to form the part. This process is appropriate for very low-volume production or for prototyping a higher-volume casting process because a new RP piece is required for each casting.

This book provides further discussion regarding the use of RP&M patterns for urethane and metal casting (see Chapters 4, 5, and 11).

An important, emerging application for RP is in the toolmaking (or mold and die) area. Industry is driven by the goal of reducing the time and cost of product development while assuring that the product and the process for manufacturing it are of high quality. More rapid product development means getting to the market faster, enabling a stronger market position with premium pricing, and/or improved market share. The importance of product development speed varies among market sectors; in the electronics industry, product life cycles are short and time-to-market is measured in weeks. For example, Sony has produced many consecutive models of the Walkman as a means of keeping up market interest and staying ahead of the competition. The toy industry also has a short product life cycle and a strong need for very rapid product development. Toys are developed in the spring and summer for all production and distribution timed to meet the holiday shopping season. The automotive industry is also competing on the time and cost to bring new products to market and has reduced the product-development cycle from more than 4 years down to 2 years at the leading companies. Medical product development also seeks to reduce time and cost; however, product life cycles are long and product-development times are restrained by regulatory approval processes.

For many products such as those noted, the time and cost of producing the production tools is a significant portion of the overall product-development time and cost. This is particularly true of products that will be produced in large volumes by automated processes (consumer electronics, toys, cars, etc.). For example, molding, casting, or stamping tools typically require several months to produce and cost tens to hundreds of thousands of dollars. Therefore, the possibility of positively impacting the time and cost of tooling production is appealing.

Figure 4 illustrates product-development time savings achieved by one company through the use of rapid tooling and other computer-based technologies. Several approaches for producing tooling based on rapid-prototyping technologies are at various stages of development. The earliest efforts were based on casting technology. The process mentioned earlier for producing metal castings can be applied to rapid tooling in the form of casting the tooling. Separately, rapid prototypes are being used in conjunction with a process known as Keltool to produce tools quickly. Keltool was developed by 3M and licensed to Keltool, Inc., which was recently acquired by 3D Systems. The Keltool process enables the reproduction of a physical part in metal. Within the context of rapid toolmaking, an RP model of the tool is produced and sent

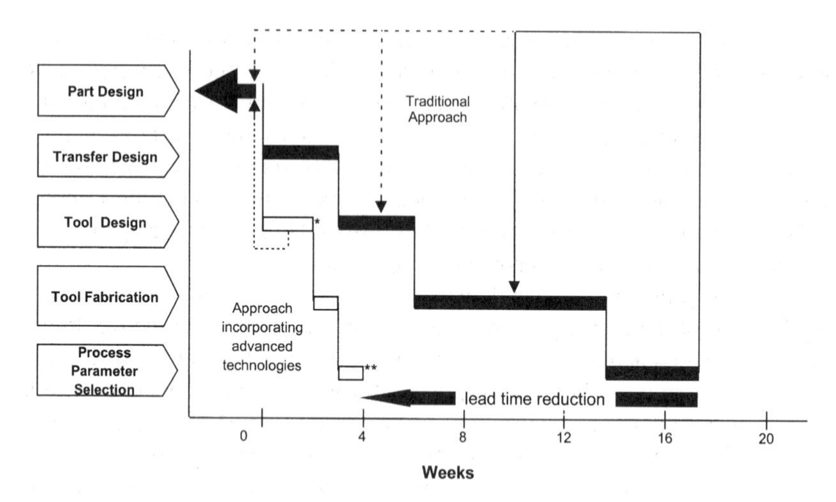


**Figure 4** Time savings with rapid tooling and other computer-based technologies. *: Includes initial process parameter selection; **: verification only.

off to Keltool for replication in metal. The Keltool process is proprietary, involving the use of metal powders to produce a composite metal piece. The Keltool approach to rapid injection-mold production has been in use at several firms for more than a year and results in the rapid production of adequate molds (see examples in Chapter 5).

More recent rapid-tooling techniques include the use of SLS to produce metal parts, a process that has been commercialized by DTM under the name RapidTool, 3D Printing of a metal tool, developed by MIT and licensed to Extrude Hone, and Nickel Ceramic Composite tooling developed by CemCom (see Chapter 6). The nickel-ceramic composite tool is created by electroforming a nickel-shell layer onto an RP part. The nickel shell is joined to a mold frame, backed with a tailored ceramic, and the RP piece is removed, creating a nickel-faced mold for injection molding of plastics. ExpressTool has more recently commercialized a production rapid-tool-making technology that also involves electroforming. The ExpressTool mold has an outer surface of nickel to achieve the needed surface hardness and durability. It is backed by electroformed copper to accelerate heat transfer, and it includes conformal cooling channels. The result is a mold that enables high-volume production with rapid injection cycle times (see Chapter 8). As mentioned earlier, Sandia is working on a one-step process for making metal parts, which is also potentially applicable to tooling.

Each of these processes has the advantage of lower costs and times when compared to traditional hard-tooling processes; however, these processes have limitations as well. The limitations are in two areas: dimensional control and long-term performance of the resulting tools. Dimensional control limitations may mean that some postprocess machining will be required, which adds time and cost. Durability limitations may make some of these tools most appropriate to lower-volume production applications and to prototyping by the manufacturing process, for higher-volume applications.

Rapid-prototyping technologies are of interest to the automotive industry because of their ability to create early part prototypes to visualize design concepts directly and for their contribution to prototype tooling to enable more substantial prototypes later in the process for testing and evaluation, as well as to test the final manufacturing processes (see Chapter 9). In the medical products industry, manufacturing of orthopedic implants, rapid prototyping is used for visualization and to check out dimensional fits, and as a means of low-volume production of cast components (Chapter 10). The aerospace engine industry also uses investment-casting technology to produce low-volume complex parts. Again, rapid prototyping provides advantage in prototyping casting processes (Chapter 11).

## REFERENCES

1. PF Jacobs. Rapid Prototyping and Manufacturing: Fundamentals of Stereolithography. Dearborn MI: SME, 1992.
2. PF Jacobs. Stereolithography and Other RP&M Technologies. Dearborn, MI: SME, 1996.

# 2
# Process Modeling

**Georges Salloum**
*National Research Council of Canada*
*London, Ontario, Canada*

## I. INTRODUCTION

Three facets of product and process optimization involve the simulation of structural behavior, material flow, and solidification. Simulation technology is used to improve part and mold design and for the optimization of die casting, injection molding, blow molding, and thermoforming operations. Such net shape material processing techniques, molding, forming, and casting are vital to the mass production of single or integrated components for numerous industry sectors such as automotive, packaging, appliances, electronics, telecommunications, medical, leisure, and sports. The success of these industries in responding to rapidly changing customer demands will depend on the ability to develop and apply state-of-the-art technology in collaboration with other partners. Instead of the traditional method, where product and manufacturing engineering follow in a sequential order, the emphasis is put on developing the product and its production process concurrently.

As a result of market globalization, the appearance of high-performance materials, increased product complexity, and geographical variation of raw material and labor costs, more pressure is put on the material processing industries and end-product manufacturers for constant innovation and process optimization. Their customers' demand for higher quality standards and lower costs often presents a considerable challenge beyond the reach of the individual companies. Despite the introduction of tools like computers, advanced processing machinery, and programmable controllers, industry continues to be plagued with problems such as voids, surface defects, flashes, cracks, warped parts, material degradation, wrinkles, specks, parts out of dimensional specification, and late delivery.

There are approximately 75,000 injection and blow molders, thermoformers, and die casters and over 10,000 die and mold makers serving a variety of industries and original equipment manufacturers. The processing of metal alloys and polymeric materials is characterized by the complex interactions among the material (resins, metal alloys, and composites), equipment configuration including mold and die design, and processing conditions. These interactions ultimately determine the processability of the material, the economics of the process, and the properties of the final product. Therefore, it is essential to develop a unified approach which incorporates simulation models for the material behavior, processing operation, and product performance.

## II. CONCURRENT PRODUCT AND PROCESS DEVELOPMENT

Figure 1 shows how a concurrent manufacturing approach can be applied to the development of new products and processes. The first stage is to plan the process and define the specifications for the part, mold, and machinery which will be used during the development cycle or for production purpose. It is very important technically and economically to select the best material, various tool steels, stainless steels, ceramics, graphite, and nonferrous alloys such as Al, Cu–Be and Ni, during the construction of the mold. Also, soft tooling materials including Bi–Sn, epoxy, urethane, RTV rubber, and kirksite compounds are commonly used to produce composite molds for rapid prototyping.

More recently, nickel-, cobalt-, and tungsten-based superalloys are being used for making inserts, cores, runners, and gates to sustain high melt temperature and abrasive fillers. High speed and laser machining, electrical discharge machining (EDM), WEDM, vacuum-assisted casting, rapid tooling, and prototyping technologies such as stereolithography (SL), selective laser sentering (SLS), laminated object manufacturing (LOM), solid ground curing (SGC), 3D processing, lost foam casting, thermal spraying, NVD and electro-chemical deposition are available to speed up the development process.

The design of the mold and the selected materials have a direct impact on the fabrication methods to be used. Two-plate and three-plate molds with or without stripper or rotary plates, insert molds, in-mold decoration, stack

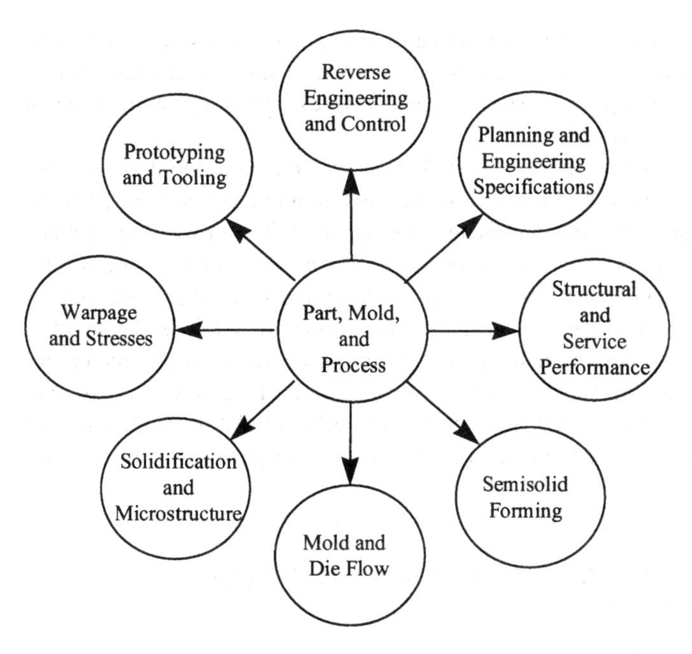


**Figure 1** Integrated process/product development cycle.

molds, parting line molds, hot runner molds, multicavity molds, cam action and multislide molds, lost core molds, and collapsible core molds are examples of the potential degrees of freedom available to the designers and mold makers. There are several types of processing machinery to be considered during the product-process development stage such as vertical and horixontal presses with or without rotary and shuttle tables.

Concerning the choice of the material for the product itself, there are a great number of polymers, metal alloys, and composites to be evaluated. Commodity, engineering, and high-performance materials are available in specific or customized compositions for a given application or market. A material demand profile should be prepared taking into consideration the structural, functional, environmental, processing, and recycling requirements. Three-dimensional computer-aided design/computer-aided manufacturing (CAD/CAM) systems operating on workstations and personal computers are available to assist the designers and process engineers. They can access various commercial computerized databases provided by the material suppliers for the selection of the most suitable material.

The designer can apply computer simulation to conduct structural analysis, evaluate part functionality, and verify that the part stands up to the demands that will be encountered under service conditions. Beyond this, experienced designers follow accepted practices which directly influence the weight and strength of parts. They know it is important to avoid sharp corners and maintain relatively constant wall thickness to minimize stress concentrations.

As the part design evolves, it is also important to evaluate how various design options or alternatives influence how the part can be manufactured. Some basic ground rules with respect to the number, location, and size of gates and draft angles should be used to facilitate the filling and the ejection of the part from the mold. However, to produce parts of high quality, the product designer or the original equipment manufacturer must cooperate closely with the mold maker and the material supplier. At the same time, the quality of a part in terms of strength, appearance, and dimensional tolerances is directly dependent on how it is produced. It is crucial to recognize that even a perfectly designed part can be ruined and fail specifications if it is formed under inappropriate conditions.

## III. FINITE-ELEMENT MODELING AND SIMULATION

Material processing generally involves the transformation from a solid state, usually in the shape of ingots, pellets, or powder, unreinforced or reinforced, through a liquid phase into a final solid product with a specific shape, dimensions, and properties. These phase transformations may involve several steps: heating and/or melting, forming, solidification, and finishing. During processing operations, the material experiences simultaneous fluid flow (laminar or turbulent) and transient heat transfer (conduction, convection, and radiation). Flow regimes, depending on the nature of the material, the equipment, and the processing conditions, involve combinations of shear and extensional flows in conjunction with enclosed-surface or free-surface flows.

As shown in Fig. 1, the optimization of the process and the product must be based on a very good understanding of the interactions between the material behavior during flow and solidification, and structural deformation. In general, computer simulations deal with two main aspects (i.e., mathematical modeling of the process and numerical methods employed to find the solution). The analysis of material processing operations involves the solution of the equations of conservation of mass, momentum, and energy. The equation of conservation of momentum represents the balance between the kinematics (velocity field) and dynamic variables (pressure and stress field) acting on any given fluid. Therefore, it is necessary to introduce an expression that relates the velocity and stress fields. This relationship, called the ''rheological constitutive equation,'' is used as the defining equation for specific types of material behavior under flow (i.e., Newtonian, viscoelastic, viscoplastic, etc).

The equation of conservation of energy represents the balance of heat transfer to and from the system due to convection, conduction, viscous dissipation, phase change, and so forth. Furthermore, in the case of compressible fluids, where density variations are important, it is necessary to employ an appropriate ''equation of state'' to describe pressure–volume–temperature (P–V–T) variations. Depending on a particular geometric configuration, the deformation may be dominated by shear or tension or may involve a combination of both. Shear-dominated flows are frequently associated with flows in closed channels or cavities having constant cross sections, whereas extensional flows often accompany deformations in certain types of free-surface flows.

The resulting system of nonlinear partial differential equations representing the phenomena taking place during processing can only be solved using numerical techniques such as the finite-element method. The use of numerical techniques has also gained popularity due to the development of solid CAD models capable of representing three-dimensional objects of complex geometry with automatic mesh generation based on topological searching and adaptive control algorithms.

The flow, solidification, and structural models are interrelated because it is necessary to have information regarding the thermomechanical history in order to predict the microstructure and the final properties. This involves solving the conservation equations with fewer restrictions; therefore, the complexity of the problem as well as the computational time will increase significantly. Finally, one can distinguish two-dimensional (or membrane approximation) and three-dimensional approaches that can provide the following capabilities:

1. Information on molding and casting characteristics, as well as the thermomechanical history (temperature, pressure, deformations, velocity profiles, etc.) experienced by the material during processing
2. Information on product quality and microstructure (distribution of density and crystallinity, part weight, wall thickness, wrinkles, residual stresses, shrinkage variation, porosity, orientation, permeability, dendrite arm-spacing, grain size, etc.)
3. Information on processing parameters, dies, and molds such as runners, risers, and gates, cooling channels and ejector pin layout, injection speed and pressure profiles, clamping and ejection forces, and so forth

## IV. INJECTION-MOLDING AND DIE-CASTING PROCESSES

Injection molding and die casting are the most important processes for the production of three-dimensional parts from plastics, metal alloys, elastomers, composites, and, increasingly, metal and ceramic powder. These processes involve the mixing, melting, and/or heating of a material followed by its injection, under pressure, into a mold where it subsequently solidifies. The injection step involves the delivery of the melt from the tip of the cylinder or screw through the sprue, runners, and gate into the cavity. During the filling stage, the material flows into the mold under pressure or by gravity casting.

The filling stage is concerned with the transient nonisothermal flow (laminar or turbulent) of the fluid. When the cavity is filled, a high-packing/intensification pressure is applied to force additional material into the mold. The purpose of this step is to introduce sufficient material into the cavity in order to compensate for shrinkage during the solidification stage and to ensure proper filling of complex parts. The increase of the material density in the cavity is responsible for the rapid increase of the pressure during this stage.

Simulation models are useful for the prediction of the filling pattern, short shots, voids and weld surfaces, pressure, velocity and temperature distributions, and the overall cycle time. Figures 2 and 3 show the predicted filling pattern during the injection molding and die casting of two complex parts. In the solidification stage, a continuous decrease in cavity pressure is observed.

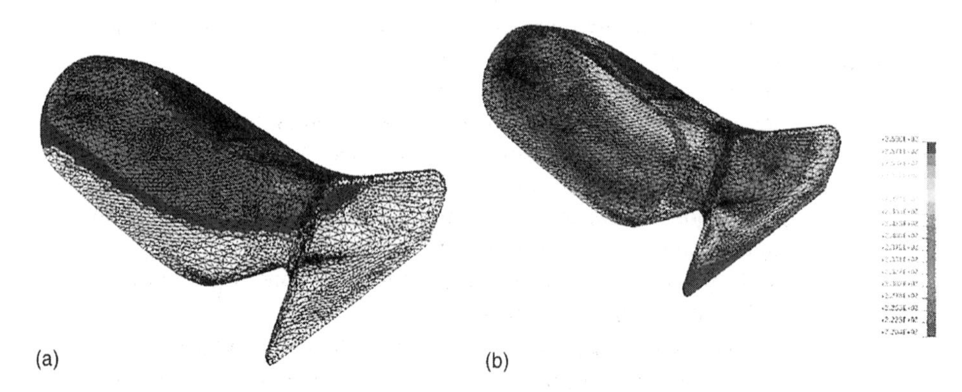


**Figure 2** Injection-molding simulation of a car mirror holder: (a) flow front at 60% of filling; (b) temperature distribution (°C).

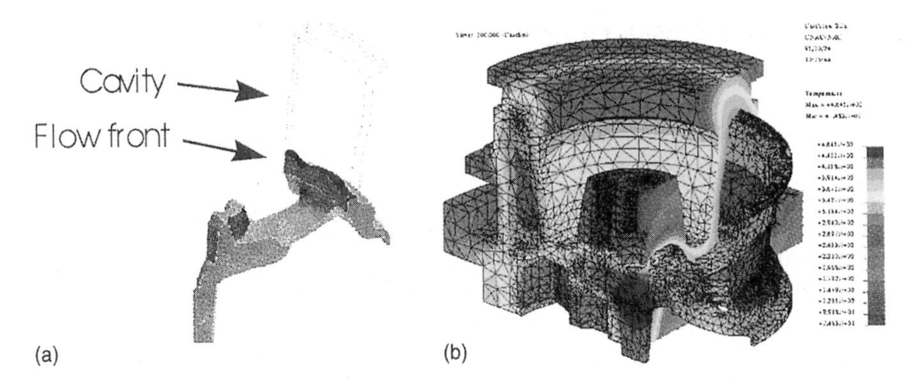


**Figure 3** Low-pressure casting simulation of a car wheel: (a) flow front during 3D filling; (b) temperature distribution in the mold and part (°C).

Cooling is continued until the solidified material is rigid enough to be removed from the mold without damage.

These finite-element simulations have also proven to be of great value in the design optimization of part geometry, to determine the dimensions of mold cavity and core and to control warpage and shrinkage while minimizing process-induced residual stresses and deformations. The thermomechanical history experienced by the material during filling and solidification has a great influence on the structural behavior of the part. These finite-element methods are applicable to the following casting and molding operations (1–3):

- Sequential injection molding to relocate weld surfaces and to minimize warpage and residual stresses.
- Coinjection molding for the production of multilayered parts offering functional characteristics in the inner or outer layers or to permit the use of recycled resins and cellular plastics
- Gas-assisted, lost core, and multishell injection molding for the production of hollow cross sections in the molded parts such as air-intake manifolds, valve boxes, and pump housings,
- High-pressure and semisolid die casting of A1 and Mg components with thin-wall or complex geometry
- Low-pressure, counterpressure, and permanent mold or gravity casting operations of thick wall parts which may require sand cores.
- Powder injection molding of metal and ceramic composites and superalloys.

## V. BLOW-MOLDING AND THERMOFORMING PROCESSES

Blow molding is a commercially important polymer processing operation used in the manufacture of hollow plastic articles. Bottles, containers, automotive, and appliances represent the most important markets for blow-molded plastic products. Blow-molding applications are expanding as a result of the success of multilayer extrusion processes. This growth runs parallel with the development of new high-performance materials suited for the three basic variations of the process: extrusion blow molding, injection blow molding, and stretch blow molding.

In the extrusion blow-molding process, the raw material is fed to an extruder in granular or pellet form. The molten material is extruded through an annular die either continuously or intermittently in order to produce a hollow cylindrical tube called a parison. Once a parison of the desired length has been formed, the mold is closed and the parison is inflated to fill the shape of the mold cavity by internal air introduced through the die-head assembly. The part is then cooled, solidified, and ejected from the mold. Figure 4 shows the parison deformation and thickness distribution during the clamping and inflation stages.

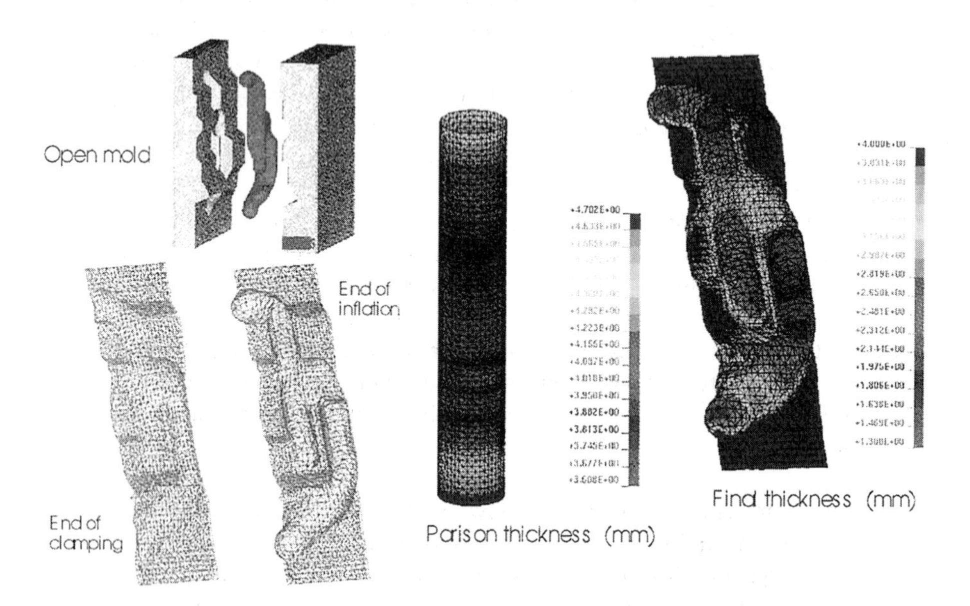


**Figure 4** Extrusion blow-molding simulation of a car plenum.

In the modeling of this process, the first objective is the prediction of the wall-thickness distribution of the molded part and the stretch ratio of the inflated parison. It is now possible to relate the predicted wall-thickness distribution to the programmable controller for setting the die gap during the extrusion of the parison and for designing the mold cavity and movable inserts or cores. The rheological complexity of predicting parison behavior arises because this is a free-surface problem involving time-dependent, nonisothermal elastic recovery from the flow of a viscoelastic fluid subject to gravity.

Injection blow molding is a process in which a preform is injection molded around a blowing mandrel. The molded preform is later preheated or rotated instantly to a blowing station having a split mold with the desired shape. The polymer is then inflated and solidified in much the same manner as in the extrusion blow-molding process. An important variation of either extrusion or injection blow molding is the so-called stretch blow-molding process. In this process, the parison or the preform is mechanincally stretched before the inflation stage. The result is a lighter product biaxially oriented for better mechanical and optical properties.

Thermoforming is the process of shaping a heated thermoplastic or composite sheet by applying either positive air pressure, a vacuum, mechanical drawing, or combinations of these operations. The objective of computer simulation of thermoforming is the accurate determination of thickness distribution throughout the final part. This is of great importance in the fabrication of complex three-dimensional parts. In such a situation, it is possible to have wrinkles, surface defects, and holes at corners or unacceptable thinning in other highly stretched areas. The prediction of thickness distribution via computer enables the designer and process engineer to select an optimum from many possible alternatives.

A finite-element analysis based on the membrane approximation is applicable to thin-walled parts where the bending resistance can be negligible. In the case of thick-walled parts, a three-dimensional formulation is required or the membrane approximation has to be relaxed in order to take into consideration the presence of compressive and bending stresses. Figure 5 shows the predicted sheet deformation and thickness distribution for a scanner cover. The methodology is applicable to the following blow-molding and thermoforming operations (4–6):

- Extrusion and coextrusion blow molding for the production of containers and multilayered hollow products.
- Sequential extrusion and 3D blow-molding operations for the production of multifunctional and flashless parts primarily for the automotive and other transportation industries

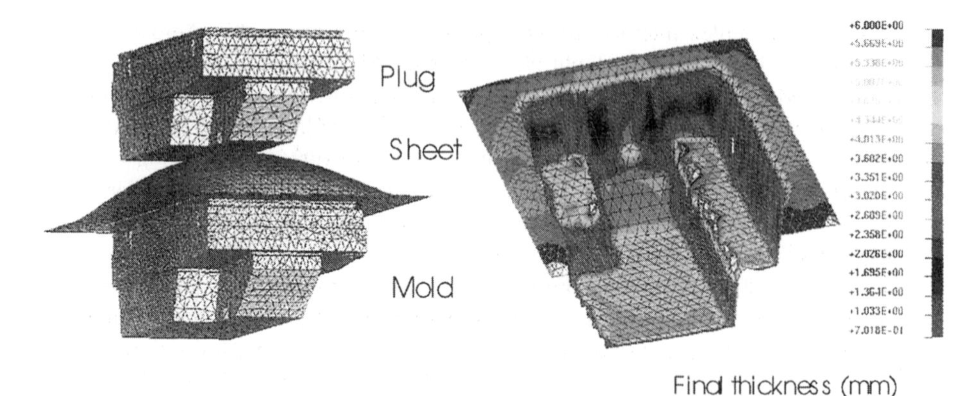


**Figure 5** Plug-assisted thermoforming simulation of a scanner cover.

- Injection and injection stretch blow molding with accurate dimensional control of bottles for the food, medical, and industrial packaging industries.
- Vaccum and plug-assisted forming with applications in food and medical packaging such as formed trays and blister packages
- Pressure and twin-sheet thermoforming with applications in panels and control cabinets, household, and consumer products.
- Drape and matched mold forming in automotive and industrial packaging

## VI. CONCLUSIONS

Numerical simulation techniques that account for heat transfer, fluid flow, phase transformations, and stress deformation are generic and beneficial to processor, equipment, and material suppliers and to end-product manufacturers. The only practical means to achieve rapid tooling and the production of high-quality parts is through the integration of part and mold design with process development. Concurrent process modeling, design analysis, and optimization will improve product performance, prevent or control processing defects, and shorten production time.

The interplay of part design, tool design, material properties, production conditions, and part quality is extremely complex and involves a matrix of many variables. It is not reasonable to expect a team of engineers to deal with these complex interactions and optimize casting and molding processes without the use of modern finite element methods.

In addition to modeling and simulation techniques, once the first prototypes have been produced, other performance tests should be conducted under service conditions. Among the tests most relevant to various applications are the evaluation of chemical and impact resistance, clarity, substance absorption and degradation, permeability, microcracking, delamination, surface roughness, discoloration, and so forth.

Computerized process simulation can be used to monitor the influence of design alternatives on processability of the part and to select operating conditions that assure the required part quality. It is clear that finite-element simulation increases process and product reliability.

## REFERENCES

1. J-F Hétu, DM Gao, A Garcia-Rejon, G Salloum. 3D finite element method for the simulation of the filling stage in injection moulding. Polym Eng Sci (in press).
2. KK Kabanemi, H Vaillancourt, H Wang, G Salloum. Residual stresses shrinkage and warpage of complex injection molded products: Numerical simulation and experimental validation. Polym Eng Sci 38(1):1997.
3. CA Loong, S Bergeron, DM Gao, J-F Hétu. Resolving die design and manufacturing problems using an integrated computer software package. Compte-rendu de la conférence ''15th International Diecasting Conference,'' Montreaux, Switzerland, 1996.
4. D Laroche, RW DiRaddo, R Aubert, A Bardetti. Process modelling of complex blow moulded parts. Plast Eng December 1996.
5. D Laroche, RW DiRaddo, L Pecora. Closed-loop optimization and integrated numerical analysis of the blow moulding process. Proceedings Numiform '95–5th International Conference on Numerical Methods in Industrial Forming Processes, Ithaca, NY, 1995.
6. ME Ryan, MJ Stephenson, D Laroche, A Garcia-Rejon. Experimental and theoretical study of the thermoforming process. American Institute of Chemical Engineers (AICHE)/Polymer Processing Society (PPS) Joint Meeting. Chicago, 1996.

# 3

# Rapid Product Development

**Paul F. Jacobs**
*Laser Fare—Advanced Technology Group*
*Warwick, Rhode Island*

## I. INTRODUCTION

In 1999 over 99.99% of all injection-molded plastic parts manufactured throughout the world will be created by tools that were either (a) machined, (b) formed by electrical discharge machining (EDM), or (c) generated by some combination of these methods. Production tooling is typically fabricated from steel, with aluminum used for molding smaller quantities. Machining was formerly done manually, with a toolmaker checking each cut. This process became more automated with the growth and widespread use of computer numerically controlled or CNC machining. Setup time has also been significantly reduced through the use of special software capable of generating cutter paths *directly* from a computer-aided design (CAD) data file.

Spindle speeds as high as 100,000 rpm provide further advances in high-speed machining. Cutting materials such as cubic boron nitride, which approach the hardness of diamond while possessing outstanding thermal conductivity, have demonstrated phenomenal performance *without* the use of any cutting/coolant fluid whatsoever. As a result, the process of machining complex cores and cavities has been accelerated.

The good news is that the time it takes to generate a tool is constantly being reduced. The bad news is that even with all these advances, tooling can still take a long time and can be extremely expensive. Six months and $250,000 is not uncommon for a large, highly detailed tool involving numer-

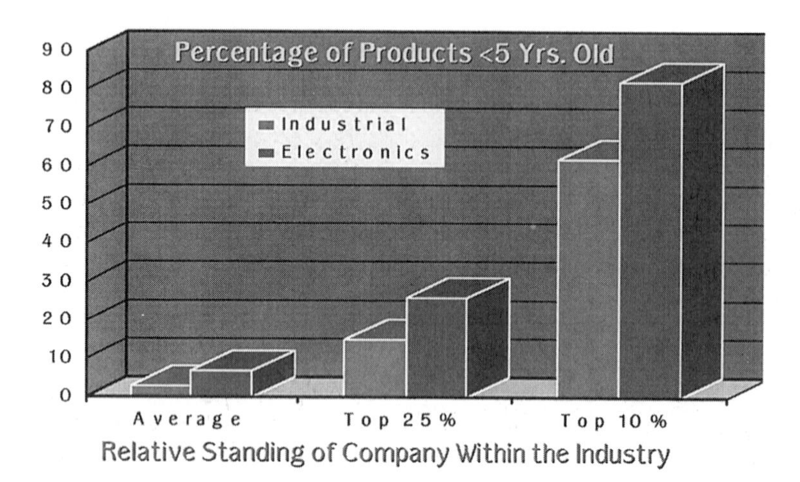


**Figure 1** The importance of new products.

ous narrow slits, high-aspect-ratio bosses, shutoffs, and multiple slide actions. Three months and $30,000 is routine for a small mold with only moderate complexity.

Many executives now realize how vital it is to move new products to market rapidly. A company able to launch a quality product ahead of their competition not only realizes 100% of the market *before* rival products arrive but also tends to maintain a dominant position for a few years even *after* competitive products have finally been announced (1). For most products, these two advantages are dramatic. However, when a new version of a laptop computer has an effective product life of only 18 months, being first to market can be critical.

Rapid Product Development is now a key aspect of competitive success. Figure 1 shows that only 3–7% of the product mix from the *average* industrial or electronics company is *less than* 5 years old. For companies in the top quartile, the number increases to 15–25%. For world-class firms, it is 60–80% (2). The best companies continuously develop new products. At Hewlett-Packard, *over 80% of the profits result from products less than 2 years old*! (3).

## II. THE WIDGET: A STORY ABOUT TIME

Let us consider a typical product development cycle for a Widget. One Friday afternoon in April, a senior engineer named John suddenly gets an idea for a new type of device that he believes people really *need, want* and *will buy*. He then spends a few hours sketching the fundamental characteristics of the idea while jotting down some notes about potential features and benefits. Just after John thinks "this is a great idea," he also wants to get a co-worker's opinion. However, he does not want to look silly. Realizing that there is not enough detail at this point, John decides to take the idea home and "sleep on it." If it is really such a great idea, it will still be a great idea the next morning!

However, the next morning dawns cold and gray. While showering, John thinks: "How many people will actually buy a Widget? At what price? How much should a Widget cost to make? How would one make a Widget? What materials should one use? What process?"

When John arrives at work the following Monday, he is both elated and frustrated; elated that he has the kernel of an idea that could be really significant, and frustrated that there are so many unanswered questions. He needs some feedback. So, John talks to Harry. Harry is absolutely convinced that nobody needs a Widget. Every time John tries to explain how terrific a Widget would be, Harry presents three reasons why it will never work, why it will cost too much to make, and besides, he heard that some company was already working on something like that.

Deflated, but not totally crushed, John develops the idea for another 2 weeks. He prepares more detailed sketches, thinks about key functions, forms "ballpark estimates" of what it might cost to make a Widget, comes up with an improved version of the idea, and sharpens his arguments in preparation for further discussions. This time John talks with Nancy. Nancy thinks that the general idea of a Widget is great, but she has no idea how large the market might be. She does believe that whatever the market is, it is likely to be price sensitive. Nancy feels that "since people have never had Widgets before, they clearly are not necessities; rather they fall into the 'nice-to-have-but-not-essential' category."

Furthermore, Nancy thinks that the Widget should be mostly made out of plastic to keep the price down. Because the intended use is very demanding, it must also be tough, so something like glass-filled polycarbonate is probably appropriate. Nancy's final recommendation is that the idea is sufficiently interesting that it would be worth having a meeting with key people from Marketing, Product Design, Engineering, Production, and Sales.

George, who is the VP of Marketing, happens to be on vacation. Edward, the VP of Sales, is at a convention in Boston. The earliest possible time for the initial meeting is the following Monday. Note that over 3 weeks have elapsed "After Concept Germination" or ACG until the first meeting is held to even *discuss* the topic.

The meeting, scheduled to start at 9:00 AM, actually starts at 9:14 because Bill, from Product Design, was working on a change to another product that is now behind schedule. Laura and Andrew, from Engineering, need to leave at 9:55 because they must attend a critical quarterly review at 10:00 sharp. With allowance for coffee and a statement of why everyone is here, John has only 37 min to describe what a Widget is, how it would basically work, what are its benefits, why people would buy one, and roughly what one would look like.

George and Ed do not understand the concept drawings, whereas Bill immediately recognizes that, as designed, the Widget would be nearly impossible to build at a reasonable cost. Richard, from Production, agrees that some design changes will be needed to simplify the manufacturing process. Laura thinks the basic idea is good but is concerned about potential thermal problems, and Andrew is already developing variations in his mind. There is not enough time to establish a consensus, but George agrees that Jennifer from Marketing would be a good person to look into the potential Widget market. Bill will develop an initial CAD model from John's sketches and run them past Richard. Laura says that she can look into the thermal issues. They all agree to a second meeting next Monday.

At the second meeting, George explains that Jennifer was working on a critical project for Division B and, unfortunately, could only assemble very fragmentary information regarding the potential Widget market. Ed was unexpectedly called to a sales meeting in Denver to establish booking targets for QIII and could not attend. Laura started a thermal finite-element analysis (T-FEA) but realized that she did not have critical dimensions or material property data and could not proceed without further information. Andrew presented some concerns regarding excessive deflections due to large bending moments. Bill's CAD design had been started but got bogged down when he could not interpret one aspect of the drawing and three phone calls to John only resulted in playing telephone tag. Note that 1 month has passed and we now have a quasi-CAD design, a fragmentary market analysis, and some vague technical concerns.

During the week Bill meets with John, clarifies the confusing aspect of the sketches, and completes a first-level CAD design, which he forwards to Richard in time for the third meeting. Because George and Ed may have difficulty interpreting the CAD representation, John and Bill decide to send the CAD file to a local service bureau, ProtoMetrics, to have a full-size model built by a rapid prototyping and manufacturing (RP&M) system. They are not sure about cost and also realize that there is no charge number for this task, as the work on Project Widget has not been approved by Finance and Accounting. John thinks he can pay for the RP&M model from discretionary funds but realizes that he better meet with Eric, the VP of Finance and Accounting, to establish a budget, organize the project team, and assign charge numbers.

John contacts ProtoMetrics and discovers they are currently swamped with work. The earliest they will be able to deliver the part is 9 days. Consequently, the next meeting is moved to the following Thursday. George, Ed, Jennifer, John, Bill, Richard, and Andrew can make the new meeting date. However, Laura is presenting a paper entitled "A Finite Element Analysis of Conformally Cooled Tooling" at a conference in Dearborn and cannot attend.

At the fourth meeting, 6 weeks ACG, Jennifer presents data implying that there could be a significant market for Widgets. Also, Nancy's instinct that this market is likely to be price sensitive was correct. After speaking with Richard, Bill realizes that the design will need changes to reduce manufacturing costs. Ed mentions that he spoke with some of his sales team at the Denver meeting and they seemed excited about the Widget idea. Andrew points out that Laura had some concerns regarding thermal issues, but she is not here to present them and he does not feel confident that he can properly represent her ideas. He also believes that excessive bending moments may lead to distortion problems, so material properties and section thickness values may be critical. Everyone passes the RP&M model around the conference table, asks questions, and begins to get a sense of what a Widget looks like, feels like, and roughly how it would work.

The group agrees they are spending a lot of time on these tasks and that Project Widget should be formally launched. A need-to-know list is generated by John. George assigns Jennifer market assessment responsibility. Bill is completely overloaded on his current task, so he recommends that Donna, from Product Design, be assigned to the team. She is excellent and has just successfully completed work on a major project. Laura will work on the thermal analysis when she returns, and Andrew will continue to evaluate critical deflection issues.

John proposes that a regular Project Widget meeting be held each Monday at 9:00 AM. Eric assigns Susan, from Accounting, to assist John with developing a budget. They will also establish a schedule and work breakdown structure to assure that all key tasks are identified.

The following Monday, the fifth meeting takes place (now 7 weeks ACG). Susan points put that because no approved budget exists for Project Widget, it will take some time to complete a schedule, personnel loading, work breakdown structure, and program costing. In the meantime, people should minimize their involvement on the Widget effort and charge whatever time they do spend to special account number 99–007. After Susan's comments, the team decides to pause until Project Widget is formally approved by Conrad, the Division Executive VP, as well as Eric. Concerned that vital momentum is being lost, John sends a memo to management summarizing the work to date, the initial market estimates, and includes a photograph of the RP&M model. He forwards copies of the memo to all personnel on the need-to-know list.

After two additional meetings involving Conrad, Eric, Susan, and John, Project Widget is finally approved, but with a budget 20% lower than John's initial estimate. Eric's final comment, made in the hallway after the second meeting had concluded, was that John should feel particularly fortunate, as no other ''special projects'' had been approved by Conrad this year.

However, John does not feel particularly fortunate. It is now 9 weeks ACG and he has this disturbing feeling in the pit of his stomach that ''somewhere out there, someone else may also be working on their own version of a Widget.'' Furthermore, momentum, enthusiasm, and *esprit de corps* have all suffered during the 2-week wait for an approved budget. Also, he just learned that Bill has been reassigned to his former project. Although Donna may be terrific, she is utterly unfamiliar with the current design.

Ten weeks ACG, the sixth Project Widget meeting is held. Susan explains the new budget, schedule, and work breakdown structure. Donna is introduced to the group and notes that Bill gave her a copy of the current Widget CAD file as well as the RP&M model. She asks a few questions related to some of the geometric characteristics, and Andrew explains that they were required to increase stiffness and reduce deflection.

Laura hands out copies of her initial thermal analysis and notes that excessive heating may indeed occur in two locations. However, until more detailed T-FEA results are available, based on *actual thermal property data* for the proposed material, she cannot be certain about the accuracy of the predictions. Ultimately, the only way to be confident of the thermal design is to test a *true prototype*, injection molded in the intended 30% glass-filled polycarbonate. Unfortunately, this will require prototype tooling.

Jennifer presents an updated marketing analysis. Her preliminary estimate suggests that the Widget market could reach $80 million this year, $120 million next year, and $180 million in year 3. Into year 4, things become fuzzy due to potential obsolescence issues and uncertain levels of enhanced performance in the future. Nonetheless, conservative estimates indicate that the total market over a 5-year product life cycle could exceed *half a billion dollars*.

At this point, everyone in the room is excited. Donna agrees to meet with John and some local toolmakers to establish estimates of the cost and schedule for prototype tooling. George and Jennifer state that they would love to have about 200 marketing test samples to generate response from buyers at major retail outlets. Ed would also like to have about 150–200 prototypes to get some feedback from his salesmen.

During the week, Donna and John meet with three tool and die shops. They are told that ''400 prototypes is a really nasty quantity.'' If they needed only a dozen, then soft tooling using an RP&M master, silicone RTV, and two-part polyurethanes might suffice. Although the mechanical and thermal properties of various polyurethanes would *not* be identical to 30% glass-filled polycarbonate, at least the cost would be low and they could have their parts within a few weeks.

However, if it is critical that they have *true prototypes*, injection molded in glass-filled polycarbonate, then all three toolmakers suggest aluminum prototype tooling. It will cost less than steel tooling and could be ready in 12 weeks. Still, CNC-machined aluminum tooling will be difficult to amortize over only 400 prototypes. Also, if there are any additional product design *changes*, tooling rework can be expensive and will push the delivery date out even further!

Carefully inspecting the RP&M model, one of the toolmakers notices a small undercut which would require a slide action. This will further increase the cost of the prototype tooling and extend its delivery. He inquires if the design could be changed to eliminate the undercut? Donna says that she will look into a design modification, will develop a new CAD file, and also have a second RP&M model made. Donna and John leave the toolmaker realizing how important it was that this problem was detected *now* and that an iteration of the design should not be too difficult because RP&M models can be built relatively quickly and inexpensively.

By the seventh meeting, Donna has made subtle changes to the CAD design. However, George, Jennifer, and Ed are not sure what effect these alterations may have on aesthetics. The group decides to purchase *four* RP&M models of the new design, one for each toolmaker to improve communication and reduce bidding uncertainty, and one for the Widget team.

Donna develops a .STL file from her new CAD design, having discovered that this is easier for ProtoMetrics to work with and will reduce their price as well. She then forwards the .STL file to the service bureau. Unfortunately, they are still swamped with work and can only promise delivery in 7 days. John approves the purchase order, but he must now reschedule the project meeting for Wednesday. The new meeting date is exactly 3 *months* ACG.

At the eighth Project Widget meeting, Donna passes the new RP&M models around the room. The undercut has now been eliminated and the aesthetics look great. Furthermore, the prototype tooling will be simpler, less expensive, and will be able to be delivered more rapidly. After the meeting, Donna and John take one new RP&M model to each of the three tool and die shops, requesting formal quotations on the machined aluminum prototype tooling.

The following Friday, they have received all three bids. Two of the shops are quoting 12 weeks and about $50,000. The third shop is quoting 10 weeks and roughly $62,000. Although time is certainly important, it is extremely difficult to convince Susan that 2 weeks is worth $12,000 just for prototype tooling, so the team decides to go with Central Tool & Die's 12-week bid for exactly $50,176.

During the 12 weeks that the prototype tooling is being fabricated, Jennifer starts the layout of the various marketing collateral materials, including packaging design, photographs, sales brochures, detailed product specifications, health and safety compliance information, Underwriters Laboratory (UL) certification forms, advertising storyboards, and so forth.

Meanwhile, Laura completes a more detailed T-FEA and concludes that the Widget will probably be operating in a safe regime. However, the temperatures in the two anticipated ''hot spots'' remain a concern. Consequently, Laura strongly recommends that detailed thermal testing of *true functional prototypes*, injection molded from the final intended material, will be required to establish the actual safety margin, if, indeed, there *is* a safety margin.

Andrew has also completed a mechanical finite-element analysis (M-FEA) and concludes that his original concerns about the part's stiffness were indeed appropriate. There is an issue with excessive deflection causing potential interference during operation. Unfortunately, the margins are sufficiently close that only careful deflection measurements on a functional prototype will truly establish design verification. Also, the potential hot spots identified by Laura will tend to reduce the modulus of elasticity of the material, which could further increase the deflection, making the problem even worse.

Andrew notes that this is a classic example of an ''interactive effect'', where normal operation results in mutually dependent thermal and mechanical loads. Specifically, the increased temperatures in the two hot spots locally weakens the material, leading to increased deflection. Simultaneously, the increased deflection slightly alters the thermal boundary conditions, which will change the temperature distribution. The interactive effects may be quite small or they may prove to be significant, especially if the design is ''right on the edge'' of passing or failing to meet product specifications. It is precisely this sort of thing that is difficult to predict analytically and is yet another reason why the team will never have ''warm fuzzy feelings'' about the design until reliable test data have been gathered from a true prototype.

Finally, after numerous calls to Central Tool & Die, John is informed that the tool will ''only be 3 days late.'' Apparently, he should be happy about this. Because Central has some small injection-molding proof presses, they could run the first 20 parts on Friday afternoon. John could then pick them up in time for the regular Project Widget meeting on Monday. The remaining 380 parts could be run the following week, or the tool could be forwarded to a local injection-molding shop to run the rest of the parts within a day or two. John agrees to pick up the first 20 parts on Friday but decides to wait until after the Monday meeting to select the injection-molding vendor for the remaining Widget prototypes. While jotting a reminder in his calendar to visit Central Tool & Die on Friday afternoon, John happens to notice that this will occur almost exactly 6 *months* ACG.

At the next meeting, John distributes the injection-molded prototypes. Everyone is impressed with their overall look and feel, but final assembly and functional testing still remain to be accomplished. Laura and Andrew agree to start testing as soon as possible. Laura's technician, Joan, is out sick with the flu, but John agrees to help Laura assemble and calibrate the required thermocouples. Andrew has already carefully calibrated six strain gauges in preparation for mechanical testing and will apply them to a second prototype. The accumulation of real test data should begin the next day.

Within 2 days, the results of the functional testing are complete, and a special Project Widget meeting is called. First, Laura presents the results of the thermal testing. The measured temperatures are within 10°F of the T-FEA predictions and, indeed, there are two hot spots. At 120% of peak anticipated loading, the temperatures are still within specification, although the data indicate that one is nervously close to the upper allowable limit.

Next, Andrew presents the results of the mechanical testing. Here, the results are *not* especially close to the M-FEA predicted values. Indeed, the largest deflections are occurring very near the highest temperature region. Andrew strongly suspects that the elevated temperatures have reduced the modulus of the 30% glass-filled polycarbonate to a point where the stiffness is no longer sufficient to keep the maximum deflection level within specification.

This is precisely the type of interactive effect that Andrew had mentioned previously. There are two possible solutions. The first is to increase the glass loading, which will increase part stiffness and thereby reduce the maximum deflection. The second approach is to increase the section thickness, which would have a similar effect.

The good news about increasing the glass loading is that it will *not* require a redesigned part geometry. The bad news is that this will demand higher injection-molding pressure and induce a more rapid erosion of the active tool surface, thereby reducing tool life. Richard states that he has seen this kind of thing before and that in his experience, increasing the glass loading is fine up to a point, but he does not think that it will be sufficient in this case. Conversely, increasing section thickness will almost certainly work, but it would require a new CAD model, iterated T-FEA and M-FEA analyses, a modification to the tool, the generation of another set of prototypes, and yet another round of functional testing, all of which will consume additional time and money.

John agrees to call Central Tool & Die and ask them to try another 20 shots in the prototype tool, but this time using 40% glass-filled polycarbonate, which is about as high as they can reasonably go without introducing serious injection-molding issues. Meanwhile, Donna will modify the CAD design per Andrew's suggestions. Andrew agrees to help Donna with the modified geometry. Using the M-FEA program, he will perform a parametric analysis to establish how large an increase in section thickness would be required to achieve a maximum deflection within specification. It would be terrific if simply increasing glass loading will solve the problem, but everyone agrees that it would be prudent to have a backup approach as well.

The next day, Central Tool & Die shoots 20 prototypes in 40% glass-filled polycarbonate on their proof press. The parts are basically fine, except for one thin-wall section which looks a bit ragged. Central feels that this can probably be solved by increasing injection pressure and they will try that tomorrow. Meanwhile, Laura affixes her thermocouples and Andrew his strain gauges, for a second round of functional testing.

While this testing is underway, Central tries some variations on the injection pressure and determines that indeed a 10% increase seems to solve the "ragged thin wall" problem. Unfortunately, just after Central called John with this good news, Laura and Andrew bring the data from their tests into John's office. The new material has had a negligible effect on the thermal results. However, although the maximum deflection has been reduced from the prior test results, it still exceeds specification. This is *not* good news.

At this point, John calls another special Project Widget meeting. A lively discussion ensues. Many ideas are presented. Sales wonders what would happen if we only test to 100% of maximum load instead of 120%. Richard quickly states that a lot of Widgets would be broken by muscular users, the company would spend a fortune on product guarantees, and the Widget reputation would take a dive. George then asks Andrew how large a change in thickness would be required to meet the specifications. Andrew finds the latest M-FEA results from his parametric analysis and explains that only an 8% increase in section thickness should be needed.

Recognizing that this approach may be critical to the success of the entire project, George then inquires of Andrew, "What is your confidence level regarding the computational analysis?" Andrew states that M-FEA results are generally accurate to within ±10%. However, because part stiffness increases with the cube of the section thickness, if they went from 8% to just a 10% increase in section thickness, they would almost certainly gain additional safety margin.

Both John and Laura agree that increasing the section thickness by 10% is probably a good idea. Ed notes that this means that the part volume will increase slightly, as will the weight and the material cost, and wonders if any of these might be problems. Richard says that the increase in material cost will only be pennies per part. Donna says that she can calculate the increase in weight from the solid CAD model, but she does not expect it to be more than half an ounce. Jennifer indicates that although excessive weight could adversely affect the Widget market, having stronger parts that do not break under hard usage is undoubtedly far more critical to overall product success.

A group discussion follows. George expresses concern about the impact of a redesign on the Product Release Date or (PRD). Ed asks John how long he thinks it will take to come up with a revised schedule and a more accurate PRD. John mentions that the extent of the delay will depend on a series of events: How long it will take to modify the CAD design, to build a second iteration in RP&M, to bring the RP&M model to Central Tool & Die for a quote on reworking the prototype tool, the time for Central to bid, the internal approval cycle (as tool rework was not in the original budget), the actual time it will take Central to re-work the tool, shoot another 20 parts, and the time it will take Laura and Andrew to complete still another round of functional testing. Nonetheless, John states that this is really their only option, other than canceling the project. The entire team concludes that except for the deflection problem; (a) the basic design is terrific, (b) the potential Widget market is substantial, (c) they have made considerable progress, and (d) quitting after all this work and expense would be incredibly wasteful.

Collectively, a decision is made to redesign the Widget. Donna will update the CAD design, including Andrew's latest suggestions for slightly thicker walls. Richard will talk to Central Tool & Die to get an estimate of the cost and time needed for tooling rework. John and Susan will also generate up-dated costs and schedules. When this is complete, John will meet with Conrad, Eric, and Susan to secure the incremental funding. Richard will then get a firm bid and schedule from Central for the tooling rework. Andrew notes that because the redesigned prototypes will likely meet all the specifications and Project Widget will require production steel tooling later anyway, perhaps they should ask Central to prepare a formal bid for this as well. They can always go out for multiple bids later. John and Richard agree.

Within a few hours, Donna completes the CAD modifications. The critical sections in the region near the maximum deflection problem are now 10% thicker. Donna requests that Andrew check the second-iteration CAD model (i.e., without any undercut, but with 10% thicker sections). Andrew agrees that the modifications look good. He will go through another M-FEA to be sure that the maximum deflections will remain within specification.

Donna uses the solid CAD model of the second iteration to determine the weight increase relative to the first iteration. It turns out that her guess was close; the weight increase will only be 0.382 oz. Although the changes are small, Donna thinks that if the budget can handle it, they probably should have another RP&M model made. She notes it would be ironic if they were heading toward final production and everyone was touching and looking over a model of a Widget that was lighter and thinner than what they were actually going to produce. She calls John with the results and mentions the additional RP&M model. He agrees that this is a good idea and will add this to the budget increment. John also thinks to himself that Bill was right, Donna really is doing a great job.

Andrew locates some data showing the modulus of elasticity of 35% glass-filled polycarbonate as a function of temperature which looks reliable (viz. the graph contains error bars, and the test conditions are well defined). With these data the FEA predictions should be even closer to the test results. The original design called for 30% glass loading. Further, the team had already tested both 30% and 40% glass-loaded material from Central's proof press.

However, Andrew realized that 40% glass loading was probably pushing the injection-molding pressures a bit. Perhaps 35% glass loading would provide some safety margin without making life too difficult for the production molding shop. If the M-FEA data looked good, he would recommend that the final production material should be 35% glass-filled polycarbonate. Andrew believes this would be a near-optimum choice—balancing strength, stiffness, tool erosion, and ease of manufacturing.

Meanwhile, John introduces Richard to Phil, the key person at Central Tool & Die. Richard suddenly recognizes that Phil is an old friend from college whom he has not seen in years. After John updates everyone on the status of the project, he feels comfortable that Richard and Phil will deal with the tooling rework issues. John returns to his office to meet with Susan about developing the new schedule and budget.

The next day Andrew completes the M-FEA analysis on the second design iteration. As he suspected, the combination of 35% glass-filled polycarbonate and a 10% greater section thickness has reduced the maximum deflections below the product specification limits, with a nice margin of safety. He takes the results to John's office just as the senior engineer returns from Central Tool & Die. John studies the M-FEA results for a few minutes, congratulates Andrew on a job well done, and agrees that 35% glass loading is probably close to optimum.

Four days after the meeting with Phil at Central Tool & Die, Richard receives a formal quote for the tooling rework: $16,240 and 6 weeks. Richard calls Phil back, thanks him for such prompt quoting, but inquires if there is any way that the rework could be done faster. Phil informs Richard that Central has so much work at present that the dominant part of the 6 weeks actually involves queue time. Simply put, there are numerous rush jobs, and only so many machines and toolmakers, so each job basically has to wait its turn. Nonetheless, because they were old friends, Phil will do his best to try pushing their job ahead a bit whenever possible.

During the next week, John meets with Conrad, Eric, and Susan to go over the revised budget. Eric is concerned about the additional $16,000 for the tooling rework. This time, somewhat surprisingly, it is Conrad who points out that they have already spent over $500,000 on Project Widget, including burdened labor, RP&M models, prototype aluminum tooling, FEA, and functional testing. The additional $16,240 is hardly a major problem. Conrad's real concern is the cost of the production tooling, the sales and marketing collaterals, and the advertising campaign before, during, and after PRD. The modified budget is approved.

John is relieved that things went well, but cannot help think that all these extra meetings with top management require many hours of three or four very expensive people. He is also acutely aware of the irony that the final development costs for Project Widget will probably wind up very close to his original estimate. Had they not shaved 20% off the top initially, he would actually be well within budget at this point, all these meetings would not have been necessary, and he could have saved time by not having to divert his attention. Oh well, apparently some things will never change.

The next day, Donna receives the RP&M model of the second design iteration. It looks great, and the incremental weight increase is so small that it is not really noticeable. She calls John and asks him if she should check with someone in the model shop about having it painted to look just like the final design of the production Widget. John agrees and notes that if the painted RP&M model can be ready by Monday, she should bring it to the project meeting.

At the next meeting, Donna shows everyone the fully sanded, primed, and painted model. The team is thrilled to see something that looks like a product they could imagine people buying. Richard reports that with the approval of the budget increment, he has forwarded a purchase order to Central Tool & Die for the rework. The schedule calls for the reworked tooling to be ready in 6 weeks. Richard hopes Phil will complete the first 20 functional prototypes in about 5 weeks.

Jennifer shows the group the preliminary versions of the advertising storyboards. The team likes them, although minor format changes are suggested by Ed. The health and safety package is about 85% complete, and the certification forms for UL approval are ready to be submitted. UL testing will be scheduled once functional prototypes are available. Sales brochures will be prepared using photographs of the prototypes, after final assembly.

The next 5 weeks seem to take forever. John double checks with Laura and Andrew to be absolutely sure that all the thermocouples and strain gauges are fully calibrated and ready for functional testing as soon as the parts arrive. John also stares intently at a nondescript point on the wall while trying to think of anything else that he could possibly do now that might save time later. Precisely because the potential Widget market is significant, he expects that other people must realize this and may already be developing their own version.

Just over 8 *months* ACG, the first 20 functional prototypes in 35% glass-filled polycarbonate arrive on Wednesday afternoon from Central Tool & Die. John hands three of them to Laura and three more to Andrew. Within minutes, the parts are being prepared for functional testing. By Friday afternoon, the test results are rushed to an ad hoc Project Widget meeting called by John on short notice. Finally, the test results all meet specification, Hallelujah!

The project team is elated. The prototypes look terrific, the thermal results are better than before, and maximum deflection at 120% of design load is about 15% less than the specification. Operating the units as hard as they can, even Richard, who was a linebacker on his college football team, cannot induce enough deflection to cause interference. Ed requests 150 prototypes for his sales force, and Jennifer and George need about 200 for marketing test samples.

Richard notes that the completely assembled tool, including the ejector holes, ejector pins, ejector plate, cooling lines, registration pins, registration holes, and of course, their precious Widget core and cavity set, is very heavy and would take some time and money to deliver to another mold house. Furthermore, they would need to do all the paperwork necessary to cut a purchase order for a new subcontract. If they stay with Central, the paperwork would be easier and quicker. Also, because Central's proof press could easily handle another 350 parts, he suggests that they also complete the injection-molding task. Besides, Central Tool & Die has really worked hard to deliver the prototypes as soon as possible, and they deserve the business.

This time, John and Richard visit the three local tool and die shops that had previously bid on the prototype aluminum tooling. John shows them the latest RP&M model, pointing out that the only significant changes are the 10% increase in section thickness and the fact that 35% glass-filled polycarbonate has been selected as the production material. Because the final geometry is so close to the one they had bid on earlier, the three tool and die shops all understand the project requirements. Richard asks all three to prepare formal bid packages for production steel tooling sufficient to produce a minimum of 1.5 million parts in the first year, 3 million parts the second year, and as many as 4.5 million parts in year 3.

All three tool and die shops made it clear that this situation presents some interesting alternatives. Nine million parts over 3 years pretty much dictates the need for a multicavity tool or a very considerable budget for tooling rework. Because 35% glass-filled polycarbonate is highly abrasive, tool wear will likely be substantial. If the market projection is correct, peak production will occur in year 3. The output in that year alone would strongly suggest an eight-cavity tool. However, production the first year would only require a four-cavity tool. Richard realizes that numerous options exist, but which is the best one? If only he knew what the Widget market was really going to be like.

Phil at Central Tool & Die suggests to Richard that he could save some money up front and "hedge your market bet" by going with a four-cavity tool initially. This will almost certainly suffice for the first year. If the Widget market turns out to be less robust than expected, the four-cavity tool might even see them halfway through the second year. Conversely, if the market is booming, they will have bought some time with the first tool and can always purchase another four-cavity tool later. Obviously, Central will keep a copy of the cutter path program, so the setup charges will be much less the second time around. Also, if a year or so down the road they want to introduce a Super-Widget, involving some product redesign, they can continue to produce regular Widgets while the new tooling is being generated.

Almost as an afterthought, John also asks Phil for an estimate to injection mold another 350 Widget prototypes in 35% glass-filled polycarbonate using the existing aluminum tool. Phil checks the schedule for their injection-molding proof press, and finds it will be free in 2 days. Including setup costs, the whole job should not exceed $3000 and Central will deliver the 350 prototypes next week. John thanks Phil and tells him to expect a purchase order for the additional prototypes the next day, or the day after, at the latest.

The bids from the three tool and die shops are all in house by the following Tuesday. The 350 injection molded, 35% glass-filled polycarbonate prototypes have arrived from Central Tool & Die. At the next meeting, John notes that the two large boxes at the back of the room contain housings for 350 Widget prototypes. Richard will have all 350 sets of the required components assembled into fully functioning Widgets in about 10 days. George and Jennifer from Marketing can then pick up 200, and Ed and his team from Sales will receive the other 150. The team can also complete UL testing and apply for certification. Photographs will also be taken for the sales brochures, and Marketing and Sales can start to line up orders at key retailing outlets.

The group now turns to discussing the bids for the production tooling. All three tool and die shops have proposed multicavity tools. Based on peak anticipated injection-molding requirements during year 3, the other two shops have bid eight-cavity steel molds that will cost between $217,520 and $239,880, and will require 22 and 20 weeks, respectively. Central Tool & Die, per Phil's recent discussion with Richard, has bid a four-cavity steel tool sufficient for year 1 and at least the first half of year 2, at $153,142 and 18 weeks. Because this is such a critical decision, everyone agrees to take a copy of the bid with them and study it for 1 day. The project Widget team will hold another ad hoc meeting to select the production tooling contractor at 3:00 PM the next day, a total of 8 months, 3 weeks, and 2 days ACG.

The next day, after a brief discussion about flexibility, hedging their bets, Super-Widgets, the cost of an additional four-cavity mold in year 2, anticipated mold life with 35% glass-filled polycarbonate, the importance of saving even 2–4 weeks, and the excellent work done to date by Central, the group relatively quickly achieves consensus and selects Central Tool & Die. Jennifer is now confident that all the marketing and sales collateral material will be finished well before the production tooling arrives. With 2 weeks allowance to assemble the first few thousand Widgets, John is finally in a position just before Christmas to give everyone an accurate estimate of the PRD. It looks like it will be near May 15, or about 13 months ACG. Sales and Marketing still have much work to do, and production needs to plan the appropriate staffing level for product assembly, test, and shipping, but John feels as though his work is essentially done.

On March 20, John happens to be looking through the business section of the local newspaper when, with a transfixed gaze, he reads that ACME, Inc. has just announced a fantastic new product, which looks exasperatingly like a Widget. He reads the rest of the story with only half his mind on the words and the other half on a single exclamation . . . EIGHT LOUSY WEEKS! How could we have saved just 8 weeks, or better yet, how could we have saved 12 *weeks* and beaten ACME to the punch by a month! What could we have done? Everyone on the team worked so enthusiastically and with so much skill. Where could we have saved 12 weeks out of a total of 57? Would that have even been possible?

## III. SOME LESSONS IN TIME

The Widget story is very real. It happens nearly every day. It happens in aerospace. It happens in the automotive industry. It happens in the consumer products industry, in the medical device industry, in the electronics industry, and it probably has happened to you. As one reads such a story, the characters start to become real people, and we begin to identify with them. When they realize that someone has beaten them to the new market, there is a sense of loss mixed with frustration. We actually feel sorry for the whole team. They tried so hard. On balance, they did things pretty much the way you could imagine your group at your company developing a new product. And that is the real point; doing things "the usual way" is not going to work as well, or as often as it once did. *Simply stated, to win in today's hypercompetitive global environment, you need to do some things differently than the rest of the pack.*

Let us take a close look at the Widget story. Specifically, where could our friends have saved time? In hindsight, we are all experts, so let's dig into the entire 57-week Widget product-development cycle. First, let's try to save 8 *weeks*, or about 14%, which would at least put them in a *dead heat* with ACME, Inc. That will not result in the lion's share of the market, but it is much better than being late. Next, let's try to save 12 *weeks*, or about 21%, which would put them 1 month *ahead* of ACME. This would be better still, but a month is hardly a large margin, and with clever marketing and advertising, ACME might still secure half the market. Finally, let's see if it would have been possible to save 16 *weeks*, or about 28%, which would have *completely turned the tables* with respect to product release and market share. Notice the magnitudes we are dealing with in the current cases: **14%** time savings to essentially tie, **21%** cycle reduction to win, and **28%** product acceleration to provide an opportunity for market leadership.

Aiming toward the latter goal, let's review Project Widget with the intention of finding those tasks or procedures where time could have been saved. In each case, we will (1) identify a specific *event*, (2) examine the *result* of that event, and (3) propose a *means* by which the team could have saved time. Clearly, all remedies will not apply in all cases, but if we can generate an approach that provides substantially *more* than 28% savings in the product development cycle, then all the proposed methods will not be required anyway.

1. Harry pours ''cold water'' on John's new idea. As a result, it takes 3 weeks before they have the first meeting. New ideas are like seedlings; they are very fragile and can easily be killed by a frost. Organizations that intend to ''do some things differently than the rest of the pack'' need to recognize this fact of life and develop methods to encourage new ideas. Proposed suggestion: As soon as John has the kernel of an idea he thinks might be significant, coupled with his previously successful track record, he should feel thoroughly comfortable calling a meeting. Also, the first meeting need not include VP level people. Additionally, if some people cannot make the first meeting, that is fine. Remember, the goal is to get the idea in front of people, allow them to assess it, enhance it, modify it, or simply think about it. The sketches and drawings are still necessary, but surely this should not take 3 weeks. The probable savings—**1 week**.
2. The entire team pauses after Susan notes that there is no approved budget for Project Widget. The result is that 2 weeks are lost as the effort grinds to a virtual standstill. Proposed suggestion: *Recognize* that the organization is in the *business* of generating new products. Why does this come as such a surprise to upper management? In an organization that is fundamentally involved with new product development, management should investigate what has happened over the past 4 or 5 years. How many ''special'' projects were ultimately approved each year? For how many dollars? Because the company *intends to develop new products*, and all the good ideas certainly do not always happen *before* the annual budgets are approved, then why not hold an appropriate fraction of the annual budget in *reserve* precisely for this type of ''after the budget''concept. Once the group achieves positive consensus regarding the idea, the project leader would be able to negotiate with management, without having to put the brakes on everything else. The rest of the

team could continue working while the formal budget is approved. The savings in this case—**2 weeks**.

3. The budget is approved, but 20% lower than John's best estimate. The result is about 1 week lost when John has to go back for additional funding increments. This is classic. Somehow, the financial side rarely trusts Engineering, Product Development, or Production to prepare an accurate budget. This is like a shortstop who will not throw to the second baseman because he does not think he can pivot properly. This team will not make very many double plays and will lose some games it should have won. Who is better prepared to assess all the development tasks? The people who will actually do them, or a finance officer who probably does not know what a sprue is? If the people are good enough to be on the team, then let them play the game. Study track records. Has John historically been ''on the money'' most of the time? If so, go with his best estimate. If he has been conservative or optimistic in the past, then adjust accordingly. Besides, the company is looking at a *half billion dollar* potential market. If they are hampered by budgetary constraints and lose a week, that could translate into tens of millions of dollars of lost revenue and millions of dollars of lost profits. Balance this against ''saving'' $200,000 up front, which ultimately got spent anyway! If you are going to dive into a pool, you are going to get wet. Diving less enthusiastically will not keep you dry. Probable time savings.—**1 week**.
4. Going with the lower-cost prototype tooling bid, rather than spending about $12,000 more to save 2 weeks. A classic example of *pennywise and pound foolish*. The team is ultimately going to spend 57 weeks and just over $1 million to develop the Widget. This is equivalent to about $20,000 per week, or $40,000 for 2 weeks. Even using this simple ''linear'' reasoning, losing $40,000 to save $12,000 does not look especially wise. Further, the impact of the 2 weeks on ultimate market share could, and almost certainly will, be ''nonlinear,'' and far greater. Proposed suggestion: *pay the extra money to save the time*. An even better suggestion: Utilize *rapid bridge tooling* as discussed in Chapter 4. Minimum time savings—**2 weeks**.

Note that the first four items are essentially ''cultural.'' They involve using different operational strategies designed to save time. In this case, if all four suggestions were utilized, the total time saved would be about 6 weeks, or almost 11%. This is a good start. However, do not be deceived into thinking that this will be easy. *Changes in the way organizations do things is never trivial*. There is always the tendency to fall back on what has worked in the past. However, it is no longer the past, and the competition is starting to play smarter, and with better equipment.

5. The team opts for CNC-machined aluminum prototype tooling, unaware of advances in rapid bridge tooling. From a CAD model, an RP&M master pattern can provide composite aluminum-filled epoxy (CAFÉ) tooling. This method has been refined and improved by RP&M service bureaus over the past 4 years and is used when 20–500 prototypes are required in engineering thermoplastics. Another process is direct ACES injection molding (Direct AIM™). The core and cavity are built on a stereolithography apparatus (SLA),* using the ACES (accurate clear epoxy solid) build style. Hand finishing of the master patterns is required for both CAFÉ and Direct AIM. Unfortunately, female cavity *finishing* can take 2.5–3 times longer than *building* the master pattern on the RP&M system! The resulting core and cavity are mounted in a standard tool base [master unit die (MUD), DME, National, etc.] and subsequently operated on a plastic-injection-molding machine.

Rapid bridge tooling involves the following key steps:

1. Develop a solid CAD model of the desired part
2. Select a parting surface
3. Create a CAD model of the core, cavity, and any required slide actions
4. Build RP&M master pattern(s) [or the inserts themselves for Direct AIM]
5. Mold the core, cavity, and slide actions (for CAFÉ)
6. Assemble the core, cavity, and slide actions in a standard tool base
7. Injection mold true prototypes in a wide variety of engineering thermoplastics

---

* *Note*: The *machine* is called an SLA, but both the *process* and resulting *parts* are properly abbreviated as SL.

8. Accomplish all this within 3–5 weeks (depending on size and complexity)
9. Save 50–70% of the time required for conventional prototype tooling

Both CAFÉ and Direct AIM will be described in detail in Chapter 4.

In Direct AIM, the core and cavity are generated in the form of a thin shell. This enables the insertion of conformal cooling lines into the hollow space on the back of the core. The cooling lines are simply bent from thin-wall copper tubing. Either aluminum-filled epoxy or low-melting-point alloys of bismuth, antimony, tin, and lead can be used as a *backing material* to increase both strength and thermal conductivity (4). For simple tools, this work has been accomplished in as little as 1 week (5). For more complex geometries, 3 weeks is typical (6).

Fifty to 300 prototypes have been successfully injection molded using Direct AIM, in a wide range of engineering thermoplastics, including (a) polystyrene, (b) polyethylene, (c) polypropylene, (d) ABS, and (e) nylon, with the quantities typically being smaller for the higher-melting-point thermoplastics (7). For glass-filled plastics, it is difficult to successfully injection mold more than about 50 acceptable parts (8). However, if a Direct AIM tool had merely generated five successful parts in 35% glass-filled polycarbonate, the Project Widget team *still* could have completed the first round of tests, *without* the need for CNC-machined aluminum prototype tooling. The time saved on the initial prototype tooling phase *alone* would have been about **9 weeks**.

6. Furthermore, rapid bridge tooling provides additional benefits. The team would have discovered the undercut problem at an earlier date, when the part would have seized in a CAFÉ tool. Also, when they found the problem with excessive part deflection and completed the second CAD iteration using 10% thicker sections, ProtoMetrics very likely could have built a second CAFÉ tool within 3 weeks. Compare this with the 5 1/2 weeks it actually took for the rework of the prototype aluminum tool. Thus, an additional time savings of **2 1/2 weeks**.

Note that if *all six* suggestions had been followed, the team could have saved **17.5 weeks**, or about 30% of their *actual product-development cycle*. They would have reached PRD 2 *months ahead* of ACME, Inc., which would have had a dramatic effect on their share of the new Widget market. Further, they would have achieved all these time savings *before* they had even gotten to production tooling, which is still the single greatest product-development bottleneck!

7. The team orders production tooling. They select conventional steel tooling, machined using a combination of CNC and EDM, because the Widget has some very fine detail. This tooling required 18 weeks to fabricate, assemble, and test. Adding 2 weeks to injection mold the first lot of production parts, the total was 20 weeks. However, they might have used "rapid production tooling," such as 3D Keltool™, ExpressTool™, RapidTool™, ProMetal™, or Nickel Ceramic Composite™ (NCC) tooling. We shall discuss each of these methods further in another chapter. In this case, the production tooling could have been ready in 6–8 weeks. Conservatively assuming the longer time period and allowing the same 2 weeks for injection molding the first lot of production parts, the net time savings for this step *alone* would be **10 weeks**.

Had the team implemented all seven suggestions, the time savings could have been a phenomenal **27.5 weeks**, or almost *half of their entire Widget development cycle*. This is not "Fantasy Land." Multinational corporations, original equipment manufacturers suppliers, and RP&M service bureaus using rapid tooling are discovering time savings even greater than 50%. These dramatic reductions account for the growing interest in rapid tooling. The potential benefits are enormous. Some forward-looking organizations have joined consortiums to help them gain confidence during early process refinement (9–11). These companies know that once the techniques get past their "growing pains" and mature into standard commercial practice, anyone NOT utilizing rapid tooling will be at a serious disadvantage. Remember, as we said earlier, *to win in today's hypercompetitive global environment, you need to do some things differently than the rest of the pack.*

## REFERENCES

1. P Smith, D Reinertsen. The time-to-market race, In: *Developing Products in Half the Time*. New York: Van Nostrand Reinhold, 1991, pp. 3–13.
2. J Thompson. The total product development organization. Proceedings of the Second Asia–Pacific Rapid Product Development Conference, Brisbane, 1996.
3. R Neel. Don't stop after the prototype, Seventh International Conference on Rapid Prototyping, San Francisco, 1997.
4. W Morgan. Low melting point alloys as backing materials for Direct AIM™

plastic injection tooling. North American Stereolithography Users Group Meeting, Orlando, FL, 1997.

5. J Heath. Direct Tooling for Plastic Injection Molding. Proceedings of the SME Rapid Prototyping and Manufacturing '96 Conference, Dearborn, MI, 1996.
6. P Jacobs. Recent advances in rapid tooling from stereolithography. 3D Systems Report Number 70–270, 1996, pp. 4–7.
7. S Rahmati, P Dickens. *SL Injection Mold Tooling*. Prototyping Technology International, International Press, Surrey, U.K., 1997, pp. 172–176.
8. G Tromans. Casting and tooling applications of stereolithography at Rover Group. Proceedings of the North American Stereolithography Users Group Meeting, Orlando, FL, 1997.
9. Laboratory to Advance Industrial Prototyping (LAIP), Clemson University, Clemson, SC.
10. Rapid Prototyping & Manufacturing Institute (RPMI), Georgia Institute of Technology, Atlanta, GA.
11. LASER-engineered net shaping (LENS), Sandia National Laboratory, Albuquerque, NM.

# 4

# Rapid Soft Tooling and Rapid Bridge Tooling

**Paul F. Jacobs**
*Laser Fare—Advanced Technology Group*
*Warwick, Rhode Island*

## I. INTRODUCTION

Strictly speaking, the designation ''rapid modeling'' should have been utilized to describe the various layer-additive technologies instead of the more commonly used term ''rapid prototyping.'' In agreement with standard manufacturing terminology, a *model* is an item which conveys the general shape of something (i.e., the *form* of the object), as well as the nature of how it integrates with others as part of an assembly (i.e., the *fit* of the object). However, a model does *not* typically provide trustworthy information regarding the *function* of the final part, because a model is usually *not* made from the final production *material* and is almost never generated using the final production *method*.

By definition, a *true prototype* is an object produced *in* the intended material, *by* the final method of production. For components ultimately to be produced in metal, this might involve sand casting, investment casting, or die casting. For products to be manufactured of plastic, the most common processes are injection molding and blow molding. By this definition, *none* of the present rapid prototyping and manufacturing (RP&M) systems produce *true prototypes* directly. Their are two obvious reasons for this. First, none of the existing commercial RP&M systems can directly generate components in aluminum or 35% glass-fiber-filled polycarbonate, as our friends needed for their Widget (see Chapter 3). Second, even allowing for technological advances, it is very unlikely that any of the existing RP&M systems, or even those currently under development, will include investment-casting slurries, die-casting equipment, or plastic-injection-molding capability as part of their regular operation. Consequently, even if the various RP&M systems were able to produce an extensive array of end-use materials, true prototypes would still not be formed because, by definition, the final production *method* would *not* have been used.

The ability to perform part visualization, geometric verification, rapid iteration, and form optimization was certainly key to the early growth of the rapid prototyping industry (1). However, as noted earlier, this was actually rapid modeling. When the various RP&M systems can build masters possessing the *accuracy* and *surface finish* required for tooling, only then will they be capable of delivering true rapid prototypes in conjunction with an appropriate secondary process.

For stereolithography (SL), this has already occurred to some extent. Specifically, over 25,000 QuickCast™ patterns have already been converted into functional metal prototypes by means of investment casting (2). Also, SL accurate clear epoxy solid (ACES) masters are currently being utilized in the 3D Keltool process to develop core and cavity inserts for plastic injection molding (3). With some of the recent developments in rapid tooling, the ''M'' in RP&M, has finally started to become noteworthy.

## II. RTV MOLDING

The most widely used form of rapid tooling currently involves silicone RTV (room-temperature vulcanizing) molds. Of the roughly 300 RP&M service bureaus currently operating worldwide, about half now provide RTV soft-tooling capability (4). Some large corporations have also installed RTV equipment internally to produce exemplars of some of their latest proprietary products. In the case of the defense industry, various classified military equipment has also been evaluated in this manner. In these situations, the services of an external bureau not holding the appropriate security clearance are effectively unavailable.

The good news about RTV soft tooling is that it is *very fast* (e.g., some service bureaus can provide a first polyurethane part from an existing computer-aided design (CAD) file within 5 days). RTV soft tooling is also *substantially less expensive* than computer numerically controled (CNC)-machined aluminum tools. The bad news is that RTV soft tooling *cannot* generate true prototypes, as the process yields only vacuum-cast polyurethane objects. Furthermore, these parts cannot be produced by injection molding, because RTV molds are quite flexible and would deform significantly under the requisite injection pressures.

The RTV process begins with a master pattern in the ''positive'' form of the final part. The master can be generated by hand carving, manual machining, CNC machining, and so forth. However, to save time, master patterns are often built using RP&M techniques. Indeed, masters have been successfully generated for RTV soft tooling from (a) fused deposition modeling (FDM), (b) laminated object manufacturing (LOM), (c) selective laser sintering (SLS), (d) solid ground curing (SGC), and (e) stereolithography (SL).

Other commercial RP&M systems, such as those from Sanders, EOS GmbH, and Z-Corp. are also capable, at least in principle, of generating master patterns for RTV soft tooling. Whether patterns generated by these various technologies are indeed capable of the requisite master-pattern accuracy and repeatability has generally *not* been convincingly established. Statistical process control (SPC) data involving a range of pattern geometries will be required from each of these technologies before this author, and presumably a large number of potential users, will be satisfied.

The primary requirements for RTV soft tooling are that the master

- Shall *not* cause RTV cure *inhibition*
- Must possess the dimensional *accuracy* appropriate for the application
- Should be able to be sanded/polished to the required *surface finish*

The latter point is important, because RTV is intrinsically capable of replicating extremely fine details—down to the level of *fingerprints* left on a glass microscope slide! Tiny flaws on the master are picked up by the RTV mold and subsequently transferred to the part. Interestingly, this is both a blessing and a curse. The blessing is that RTV can, indeed, faithfully reproduce fine detail, but the curse is that great care must be exercised during surface finishing to avoid even the tiniest scratch appearing on the final molded parts.

Prior to pouring RTV, a sprue is mounted on the master, typically with a superglue. The sprue and master are wiped clean with a soft cloth moistened with isopropyl alcohol, to remove dust and fingerprints. The master and sprue are suspended in a clean wood or metal-forming box. The liquid-silicone RTV material is mixed under vacuum to eliminate air bubbles. It is then poured into the box and over the master while still under vacuum. The assembly is then placed in a low-temperature curing oven and maintained at about 50°C (122°F) for roughly 6–12 h.

The RTV curing process is exothermic. Thus, portions of the mass of curing rubber will exceed the average oven temperature. After the RTV has cured, the mold should be slowly cooled back to room temperature. The solidified RTV mold is extracted from the forming box and cut along a parting surface with an X-Acto knife or scalpel. From experience, it is best to produce a cut which is intentionally ''wavy'' near the outside of the mold, but smoother near the master. In this way, the positive and negative undulations of each portion of the RTV mold accurately register with respect to each other.

Figure 1 shows an SL ACES master, two sections of an RTV mold, and a hand-held scanner, built, formed, and vacuum molded by Accent On Design, for Compsee, Inc. Note the scalloped edges cut into the RTV mold to improve

**Figure 1** RTV soft tooling of a hand-held scanner by Accent On Design, for Compsee, Inc.

registration. Compsee estimates that the use of RTV soft tooling on this project saved approximately 75% of the time and 50% of the cost of this development relative to the conventional methods previously employed.

The Compsee scanner was molded from one of the various two-part polyurethane resins successfully vacuum cast in RTV molds. These polyurethanes provide a wide range of important mechanical properties, including (a) hardness, (b) tensile strength, (c) tensile modulus, (d) flexural strength, (e) flexural modulus, and (f) notched Izod impact resistance.

Table 1 lists these properties for three specific polyurethane resins, SG 95, SG 200 and 2170, distributed by MCP Systems. It also lists the same six properties for ABS, Nylon 6, and polypropylene (PP). These data indicate that recent advances in two-part *thermoset* polyurethane chemistry have provided some interesting alternatives to the standard engineering *thermoplastics*.

On a decidedly larger scale, Fig. 2 illustrates another example of RTV soft tooling. In this photograph, the ACES master is shown at the lower left, the scalpel-cut RTV mold sections are at the top, and three vacuum-cast SG 95 polyurethane radio/cassette/compact disk "boombox" housings are located on the right. Each of the three housings was made from SG 95, but different dyes were blended into the two-part polyurethane mix prior to vacuum molding. This enabled the generation of boombox housings in red, yellow, and black. A ballpoint pen is included near the model so that the reader can get a sense of the size of these parts.

It is important to underscore two additional aspects of RTV soft tooling. The first is that solidified silicone has *very poor thermal conductivity*. Therefore, heat transfer from these molds can be exceptionally slow. For RTV mold sizes up to roughly an 8-in. cube, 4–6 h are typically required before demolding the polyurethane part. If shorter intervals are attempted, inadequate dissi-

**Table 1** Mechanical Properties of Some Polyurethane Resins Relative to Those of Three Engineering Plastics

| Property | SG 95 | SG 200 | 2170 | ABS | Nylon 6 | PP |
|---|---|---|---|---|---|---|
| Hardness (Shore D) | 79 | 80 | 82 | 78 | 78 | 72 |
| Flexural strength (kpsi) | 7.2 | 6.9 | 9.0 | 6.3 | 4.7 | 2.9 |
| Flexural modulus (kpsi) | 396 | 391 | 495 | 361 | 284 | 183 |
| Tensile strength (kpsi) | 8.7 | 7.3 | 10.5 | 4.8 | 7.6 | 3.7 |
| Tensile modulus (kpsi) | 288 | 238 | 314 | 225 | 225 | 143 |
| Notched Izod (ft-lbs/in.) | 0.35 | 1.09 | 0.39 | 1.88 | 1.17 | 0.55 |

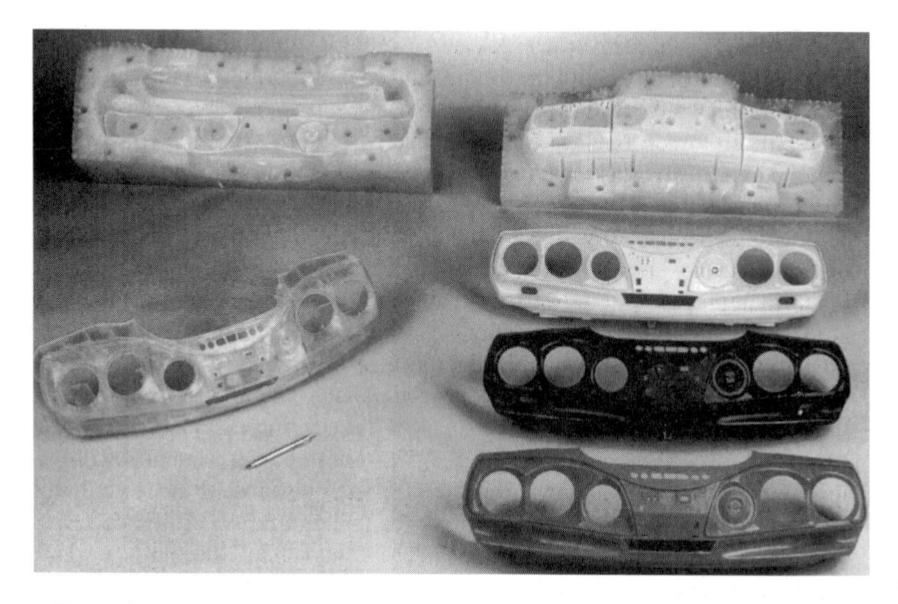

**Figure 2** RTV soft tooling of polyurethane boombox housings.

pation of the exothermic heat released during the two-part polyurethane reaction can, and often will, lead to poor part quality. For still larger RTV molds, up to 12 h may be required.

Thus, although the *generation* of the RTV soft tooling may indeed be exceptionally rapid relative to prototype aluminum tooling, the *cycle times* are certainly *not* at all fast. If only three or four parts are needed, the added time to mold and cure these parts will probably be less than 2 days. However, as many as 2–3 *additional weeks* may be consumed in simply molding and demolding 30–40 parts.

Also, RTV soft tooling is indeed soft. The good news is that the flexibility of cured silicone can greatly assist the demolding process. This is especially true of those part geometries that involve minor undercuts. In these situations, slide actions would normally be required on a conventional steel or aluminum tool. Fortunately, an RTV mold can often be sufficiently flexed to enable part release without the need to generate any slide actions, thereby greatly simplifying the mold design.

Unfortunately, the bad news is that RTV mold durability is marginal at best. For very simple geometries without sharp edges, thin walls, or high-aspect-ratio bosses, as many as 30–40 parts may be obtained from a single

RTV mold. For moderately complicated geometries with a few sharp edges, 15–30 parts of good quality can be anticipated. For highly complicated parts with numerous sharp edges, extended thin walls, and multiple high-aspect-ratio pins or bosses, only 10–15 acceptable parts can typically be produced. Beyond about 15 parts, some portion of the RTV mold is likely to be either torn or locally damaged.

Consequently, RTV soft tooling is best used when only a dozen or so parts are needed, primarily as aesthetic models for photographs or as marketing test samples. Whenever the required part quantities increase beyond about a dozen or the demands of functional testing with true prototypes become critical, RTV soft tooling is probably no longer the proper choice.

## III. INTRODUCTION TO BRIDGE TOOLING

The initial stages of project development generally involve most if not all of the following 25 tasks:

1. Concept germination
2. Initial market assessment
3. Concept refinement/definition
4. Competitive patent/legal status review
5. Generation of the initial product CAD design
6. Development of detailed functional product specifications
7. Initial thermal, mechanical, electrical, chemical, or aerodynamic analysis
8. Initial production cost/anticipated selling price estimates
9. Building a physical (possibly RP&M) model
10. Continued CAD design iteration
11. Form/aesthetic optimization
12. Initial FEA analysis
13. Identification of any potential design problems
14. Modifications to the deficient CAD design
15. Additional detailed FEA analysis/results
16. **Development of prototype tooling**
17. Generation of true prototype parts
18. Initial prototype functional tests
19. Additional CAD design changes
20. FEA analysis of the latest design

21. **Reworking the prototype tooling**
22. Generate prototypes of the new design
23. Functionally test the modified prototypes
24. Continue until all specifications are satisfied
25. Cost/Price analysis per marketing/sales inputs

Sound familiar? Our friends on Project Widget went through almost every one of the 25 listed steps in agonizing detail. Unfortunately for them, and for you, each one of these steps take *time*. An important approach to saving time and producing better products that has received a lot of attention in recent years is concurrent engineering (5).

The essential idea behind concurrent engineering is best summarized by the phrase *whenever possible, try to do things in parallel rather than in series.* This not only saves time but helps catch errors that previously ''fell into the cracks'' as a project was passed from one distinct discipline to another. Having people from each of the key disciplines work together as part of a colocated team reduces the tendency, as an example, for Laura to assume that Andrew will deal with an interface detail, and Andrew is assuming Laura will complete that task. No process developed by human beings is ever perfect, precisely because the transfer of information is never perfect. However, concurrent engineering is certainly better than whatever is in second place, and it has been documented in many studies to significantly reduce product development cycle time (6).

Nonetheless, it is hard to save a million dollars when one is only pocketing nickels and dimes. It is also difficult to effect *dramatic* product-development lead time reductions when attempting to streamline, or eliminate, tasks that may only take a few days. Surely, every little bit of time does matter, and even a small time savings on multiple tasks do add up. But a few hours saved here and a day not wasted there will rarely add up to 6 months of product development time reduction.

Rapid-time-to-market is best realized by accelerating those processes that consume the greatest amount of time! If you are going to plow a field, it is the big rocks that can ruin your plowshare, not the pebbles. Inspection of the 25 steps reveals an interesting point; 23 of the 25 tasks can typically be accomplished within about 5–10 days, with the average of them taking about 8 calendar days. If they were all done serially, the time for these 23 steps would be about 184 days or roughly 26 weeks, as shown in Fig. 3.

If most of the product-development tasks are run in parallel, it is quite likely that about 25% of the 184-day interval could have been saved. This would amount to about 6.5 weeks, as shown in Fig. 4.

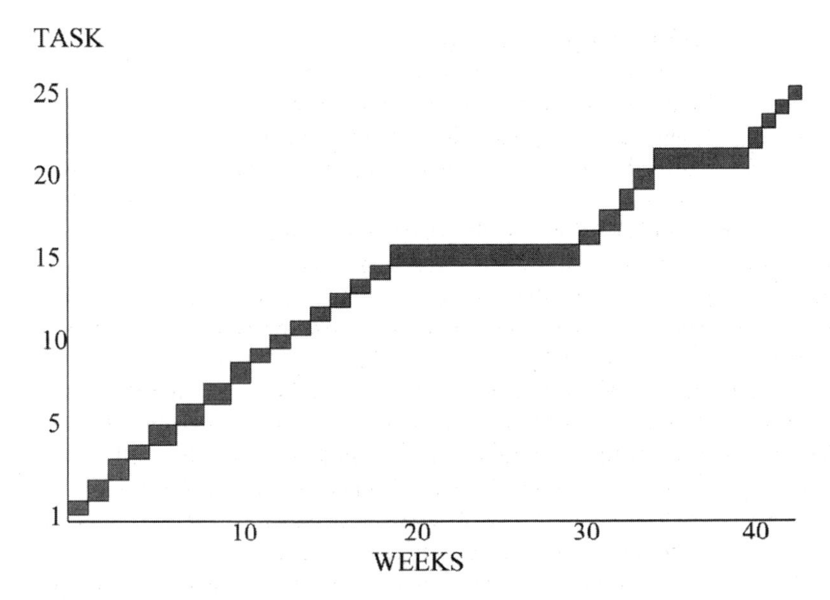


**Figure 3** A serial product-development timeline.

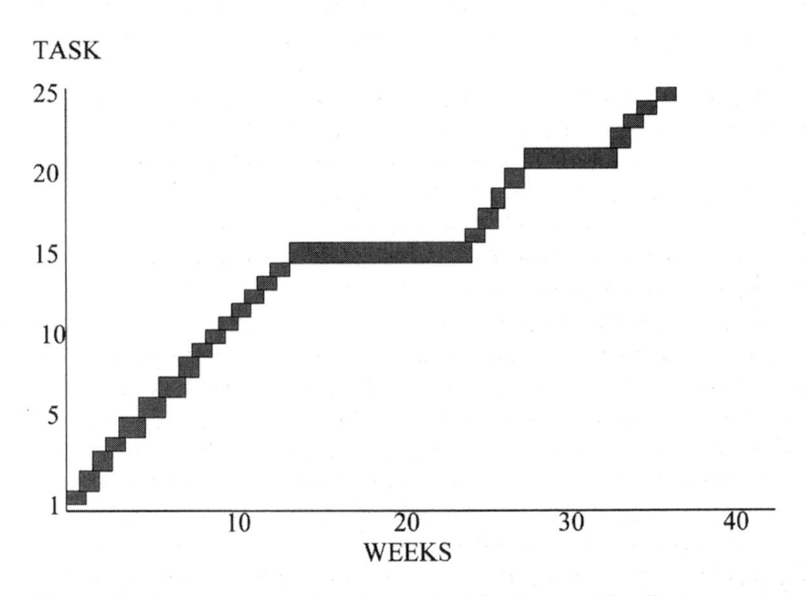


**Figure 4** A concurrent engineering product-development timeline.

As seen in Figs. 3 and 4, two key tasks, steps 16 and 21, each require far more than 5–10 days to accomplish. This is the reason these two tasks have been shown in boldface type. For Project Widget, which was intended to be representative of a ''typical'' product-development scenario, step 16, required 12 weeks to complete. Furthermore, because a problem with excessive deflection occurred, step 21 was necessary, taking an additional 5.5 weeks. The total time required for these two tasks *alone* was 17.5 weeks. Adding the 26 weeks for serial performance of the other 23 tasks, the total shown in Fig. 3 is 43.5 weeks.

Note that for this case, the development and reworking of the prototype tooling consumed 17.5/43.5, or about 40% of the time up to ordering production tooling. If concurrent engineering was used, and many of the other 23 tasks were done in parallel, the time savings would have been about 6.5 weeks, the total time prior to placing the order for production tooling would have been 37 weeks, and the prototype tooling would have consumed 17.5/37, or almost *half* the time to that point, as shown in Fig. 4.

Oh, by the way, careful review of the progress our friends made on Project Widget will show that they had completed the 25 tasks up to but not including placing the order for production tooling, in 8 *months*, 3 *weeks, and* 2 *days ACG*, or about 38 weeks! Clearly, they must have utilized concurrent engineering to a considerable extent. Unfortunately, the Widget team did *not* utilize rapid tooling.

Historically, the central problems regarding prototype tooling have been *time* and *money*. How does one produce just 50–200 parts in a production material without spending a lot of money and taking a lot of time? This has been a very real dilemma for tens of thousands of companies working on the development of millions of products. Until recently, none of the approaches were very efficient, and all were quite expensive.

Traditionally, the most common procedure involved generating aluminum prototype tooling. Although aluminum can be CNC-machined more easily, rapidly, and economically than production steel tooling, neither the time nor the money saved are enormous. If only 20 or so functional prototypes are needed for mechanical or thermal testing, it is difficult to amortize the prototype tooling over such a tiny number. Fifty thousand dollars for 20 plastic parts?

Nonetheless, many companies take the CNC/aluminum prototype tooling route. The good news is that true functional prototypes can be tested to reveal potential problems with the product. The bad news is that this step is both expensive and time-consuming, so time-to-market is further extended.

The second approach is to dispense with prototype tooling and prototypes altogether. The good news here is that considerable time and money are saved, and the PRD can be moved forward significantly. The bad news is that the product may contain flaws resulting in myriad failures, furious customers, damage to the corporate reputation, and, in some cases, even protracted and potentially onerous product liability lawsuits.

Some companies have tried a compromise approach, utilizing RTV soft tooling to quickly generate polyurethane "prototypes" that are reasonably close to the final product. The good news is this method, as discussed earlier, will definitely save time and money relative to the CNC/aluminum prototype tooling approach. The bad news is that although better than nothing, these are *not true prototypes*, and any test results based on their mechanical or thermal properties will not be fully trustworthy relative to the final product.

Fortunately, there is now a fourth option: Rapid Bridge Tooling. The term bridge tooling was chosen to suggest that this approach can "bridge the gap" between RTV soft tooling and true production tooling. *The object, simply stated, is to provide* 20–500 *injection-molded prototypes in the desired production material, quickly and inexpensively*. This is exactly what John, Laura, Andrew, and Richard needed while working on Project Widget. Had they employed rapid bridge tooling to obtain just five injection-molded prototypes in glass-filled polycarbonate, the group could have accomplished the following:

1. Completed the initial functional tests 9 *weeks sooner*
2. Discovered the excessive deflection/interference problem *much earlier*
3. Built a *second* bridge tool to *validate the increased section thickness design*
4. Shortened the program by $11\frac{1}{2}$ *weeks*
5. *Saved* about $40,000 in program cost
6. *Beaten ACME to the marketplace*!

## IV. CAFÉ BRIDGE TOOLING

Currently, there are three primary approaches to bridge tooling. The first, and most widely used, is composite aluminum-filled epoxy (CAFÉ) tooling. Many service bureaus have been generating CAFÉ bridge tools for the past few years. As an example, Laserform, Inc., located in Auburn Hills, MI (previously part of Plynetics Express) had already built more than 150 successful, water-cooled CAFÉ bridge tools (7).

From their experience, CAFÉ tools typically require between 3 and 5 weeks, have been made in sizes from 1 to 36 in., and can produce from 50 to 1000 parts, at a cost from $5000 to $20,000. Obviously, the cost and time depend on the mold size and complexity. Also, the number of parts that can be injection molded is strongly influenced by the specific thermoplastic to be molded and whether it is glass filled or not. Laserform had done up to 1000 polystyrene functional prototypes from a single CAFÉ tool, but as few as 50 parts in 40% glass-filled nylon (7).

A CAFÉ tool is typically generated directly from a positive master. The master can be made in a variety of ways, including CNC machining of aluminum, plastic, or wood. However, Laserform preferred to utilize SL masters to save time. The accuracy of SL masters is also constantly improving.

Figure 5 shows the continued reduction in the root-mean-square (RMS) error for the stereolithography process. Data for the accuracy diagnostic test

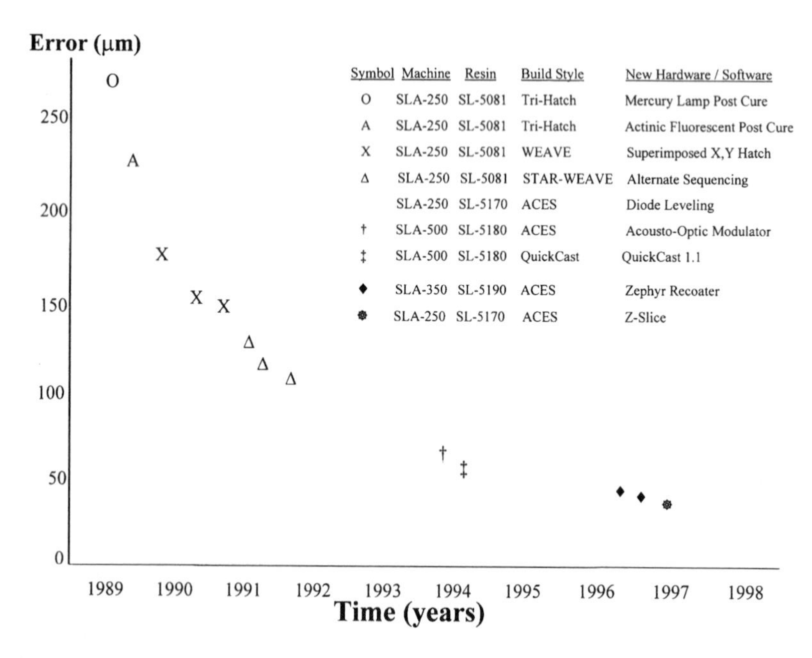


**Figure 5** Stereolithography UserPart RMS error versus time, from 1989 to 1997.

part known as the ''UserPart'' are plotted as a function of time from 1989 to 1997.

The UserPart was developed in 1989 by Dr. Edward Gargiulo, in conjunction with the North American Stereolithography User Group (8). It was designed to establish the accuracy and repeatability of the SL process. This diagnostic part measures 9.5 in. × 9.5 in. in the *X–Y* drawing plane, by 1.5 in. in the *Z* or vertical direction. For each UserPart, 78 measurements are made in the *X* direction, 78 in the *Y* direction, and 14 in the *Z* direction. The measurements range from 0.125 to 9.500 in. Each ''point'' in Fig. 5 actually represents a minimum of 1700 separate physical measurements, from at least 10 different UserParts.

It is clear from Fig. 5 that SL accuracy has improved with advances in process, resins, software, and hardware. UserPart RMS errors of a properly calibrated SLA-250, using Ciba epoxy resin SL-5170, are now under 45 μm! Pattern accuracy is one of the key prerequisites for tooling. It is not a coincidence that rapid tooling is gaining momentum as RP&M masters achieve this level of accuracy.

Building a CAFÉ tool typically starts by sanding and polishing the master pattern. This is done to eliminate ''stair-stepping'' due to the finite layer thickness used in all RP&M processes, as well as any other surface imperfections. Remember, *the surface finish of the injection-molded prototypes will only be as good as the finish on the master*. Achieving the desired surface finish is not trivial and can easily account for 20–30% of the entire time required to make a bridge tool. Currently, most of this time is spent eliminating stair-stepping artifacts. Building tooling masters with thinner layers to reduce stair-stepping requires a longer period on the RP&M system, due to the additional overhead time associated with each layer. Nonetheless, the time saved in reduced sanding is likely to be much greater. Furthermore, part accuracy will be improved, as overzealous sanding can extend below the desired CAD surface and fine features may be damaged or destroyed.

The sanded and polished master pattern is then coated with a thin film of a commercially available mold-release agent. The master is next accurately registered inside a chase box. At this point, a parting surface must be selected. If this is a plane, a simple wooden parting board can be used. However, if the parting surface is more complicated, then a machined parting board is appropriate.

Conformal cooling can be included by bending thin-wall copper tubing and locating it inside the chase box near the master. The tubing geometry can either be determined heuristically, or by means of a thermal finite-element analysis (T-FEA) performed directly on the original solid CAD model of the core or cavity. Good cooling will benefit as follows (a):

1. Reduce part distortion
2. Decrease cycle time
3. Increase productivity
4. Extend tool life

**Ultimately, conformal cooling may prove to be one of the most important benefits provided by rapid tooling.**

Next, one prepares the necessary amount of CAFÉ mold material. This involves premixing finely ground aluminum powder and two-part thermoset epoxy. The mixture is then vacuum degassed to eliminate air bubbles. Under vibration and vacuum, the CAFÉ material is poured over a master that had previously been coated with release agent. The CAFÉ mixture is then allowing to cure. Subsequently, the master and the fully cured CAFÉ mold material are inverted, the parting board is removed, additional release agent is used to coat the opposite side of the master, as well as the previously cured CAFÉ, and the process is repeated.

After about 12 h, the second batch of CAFÉ material is fully cured, the two sections are separated, and the master is removed. At this point, the core and cavity are checked for obvious flaws. If all is well, the core and cavity are aligned with registration pins/holes, ejector holes are drilled in the required locations, the ejector plate and ejector pins are installed, the conformal cooling lines are connected to quick disconnects, and, finally, the entire assembly is mounted in a standard tool base.

If the final part geometry exhibits undercuts, then sliders will be required. For simple geometries, these can be machined from aluminum. However, if the slide action involves compound curved surfaces or other intricate detail, they too can be made from CAFÉ. Obviously, as tool complexity increases so does the time required for the necessary CAFÉ tooling. Of course, the same is true for conventionally generated tooling, except that all the times involved in that case are considerably greater.

Figure 6 shows an example of a CAFÉ tool generated by Joe De Guglielmo and others at the Advanced Machining Center of Eastman Kodak. The registration pins as well as the ejector pins were machined from cylindrical steel stock, and the registration holes, as well as one insert, were also positioned within the CAFÉ core and cavity.

In this case, a specific Kodak project needed 25 *different* plastic-injection-molded geometries. Some of the CAFÉ inserts have already injection molded in excess of 1000 parts. For simple geometries, Kodak expects that

**Figure 6** CAFÉ core and cavity inserts produced by Eastman Kodak.

as many as 5000 plastic parts can be produced from a single CAFÉ tool. They have already achieved as much as 85% *lead time reduction* when employing CAFÉ bridge tools relative to conventional CNC/EDM-generated tools. In some cases, product-development cycles have been cut by a full year!

Furthermore, Kodak is typically saving about 25% in tooling cost. They are also able to rapidly (a) test, (b) iterate, (c) retest, and (d) proof multiple designs in less time that it previously took to just *test* a single design. Most importantly, Kodak can now properly evaluate *form, fit*, **and** *function* with true prototypes injection molded in the desired end-use thermoplastic.

To date, over 40 CAFÉ molds have already been constructed and operated at Kodak. According to John Fowler, Supervisor of the Plastic Development and Fabrication Model Shop, ''SL masters combined with composite aluminum-filled epoxy tooling have cut the time required for simple low-volume production molds from 8–10 weeks to just 2–4 weeks; and for complex molds from 26–38 weeks down to just 6 weeks!''

Figure 7 shows another case involving SL masters and CAFÉ tooling from Europe. ERU Elektroinstallation GmbH, in Thuringia, Germany, manufactures electrical consumer products. ERU had less than 1 year to design and test a set of universal, multicircuit, two-way-control, illuminated switches involving 27 different plastic-injection-molded components. Test results were needed prior to committing to production tooling.

Further, ERU needed the flexibility of making design modifications based on marketing inputs regarding customer preferences. Finally, it was an-

**Figure 7** CAFÉ bridge tools for ERU Elektroinstallation GmH, produced by Schilling & Partners, Engineering.

ticipated that some of these inputs/design modifications might come as late as 9 months into the program. ERU chose to work with Schilling & Partners, Engineering, a CAD, tool design, CNC, and RP&M service bureau located in Sondershausen, Germany.

First, Shilling engineers designed all 27 components in CAD. After various modifications by ERU, master patterns were generated on Shilling's SLA-250. Within just 2 months, all 27 component designs had been approved. Shilling then fabricated 27 sets of CAFÉ molds within another 3 months, or almost one every three calendar days! This reduced previous prototype tooling lead times by a remarkable 60%. Shilling was able to deliver 50 sets of all 27 components, injection molded in the desired production material, within 8 months. As a result, the various switches were submitted for VDE electrical and safety testing, and received certification a full 3 *months* ahead of schedule. In the words of Dr. Martin Schilling, ''Customers no longer talk to us about design drawings or timelines for their project's completion. Instead, we use models to communicate. Grasping an idea is much easier when you can touch it.''

Another example of the use of RP&M and CAFÉ soft tooling involves the Space Systems International division of Hamilton Sunstrand, itself part of the United Technologies Corporation. Hamilton Sunstrand is a prime contractor working on the design, development, fabrication, and assembly of portions of the International Space Station (ISS). Obviously, items designed for use on the ISS must meet very stringent requirements. However, these items will only be produced in very limited numbers. Thus, it becomes extremely difficult to justify hard tooling, as its cost cannot be efficiently amortized over such limited production quantities.

Bob Davis, Steve Irwin, and a team of engineers and scientists at Hamilton Sunstrand had to deal with this dilemma in the development of two components intended to be used aboard the ISS. Only 26 castings of each part were needed. The first component was to be produced in Inconel. Here, all 26 parts were directly investment cast using SL QuickCast™ patterns. For the second component, 26 aluminum castings were needed. The first six parts were investment cast in aluminum using QuickCast. The remaining 20 aluminum parts were investment cast using wax patterns molded in CAFÉ soft tooling. The CAFÉ core and cavity inserts were themselves generated from an SL ACES pattern. The resulting savings in time and cost were substantial relative to the use of hard tooling. Furthermore, all of the resulting 52 parts successfully conformed to a demanding 100% dimensional inspection. As a result, Hamilton Sunstrand Space Systems International is now dedicated to using the SL process and rapid soft tooling on relevant projects and is currently transferring this knowledge to other Hamilton Sunstrand divisions.

It is worth reflecting on the impact that CAFÉ tooling might have had on Project Widget. Based on results achieved by Kodak and Shilling, it is reasonable to conclude that a time savings of about 3 *months* could also have been realized by the Widget development team. Furthermore, this schedule compression occurred just through the prototype stage! Additional and very substantial time savings are possible should rapid production tooling ultimately be utilized, as discussed in Chapter 5.

## V. DIRECT AIM RAPID BRIDGE TOOLING

Another rapid bridge tooling approach involves a process known as Direct AIM™ (Direct ACES injection molding). The essential idea here is that the core and cavity inserts of a plastic-injection-molding tool are built directly on an SLA machine, using epoxy resins and the ACES (accurate clear epoxy solid) build style, discussed in detail in Ref. 10. The concept of directly injection molding thermoplastics at up to 300°C into an ACES insert fabricated from an SL photopolymer with a glass transition temperature of about 65–85°C is hardly intuitively obvious. However, after the initial results in 1995 were surprisingly positive, Direct AIM began to be evaluated in greater detail.

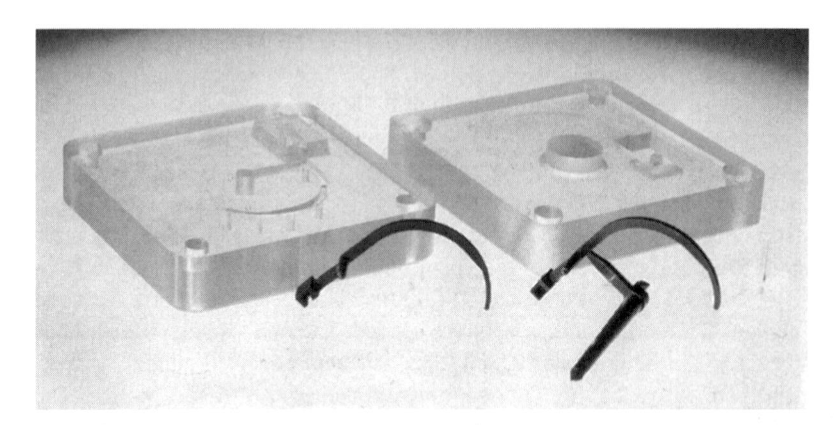

**Figure 8** Direct AIM core and cavity inserts used by Xerox Corp., and two injection-molded polystyrene switch actuators.

Figure 8 shows a Direct AIM core insert on the left and a corresponding cavity insert on the right. These inserts were built by Xerox Corporation on an SLA-250 with Cibatool™ SL 5170 epoxy resin using the ACES build style. An internal Xerox customer required 100 polystyrene switch actuators in a very short time. After evaluating various alternatives, Jeff Heath decided to try Direct AIM. His team was able to injection mold the required 100 polystyrene parts just 5 *days* after the CAD design was completed!

Table 2 provides injection temperatures, pressures, and cycle times for a number of important engineering thermoplastics that have been injection molded in Direct AIM molds. The parameters have not been fully optimized, but they have been used successfully by a number of practitioners.

Figure 9 is a so-called ''scatter diagram'' which plots measured data for a key dimension (viz. a diameter) on 200 polystyrene parts injection molded into a Direct AIM core and cavity. The inserts were held in a standard master unit die (MUD) frame.

**Table 2** Suggested Injection Molding Parameters for Use with Direct AIM Core and Cavity Inserts

| Parameter | LDPE | HDPE | PS | PP | ABS |
|---|---|---|---|---|---|
| Injection pressure (psi) | 1600 | 2300 | 2400 | 1900 | 3200 |
| Injection temperature (°C) | 180 | 220 | 200 | 205 | 240 |
| Cycle time (min) | 3.5 | 4.5 | 4.0 | 4.0 | 5.0 |

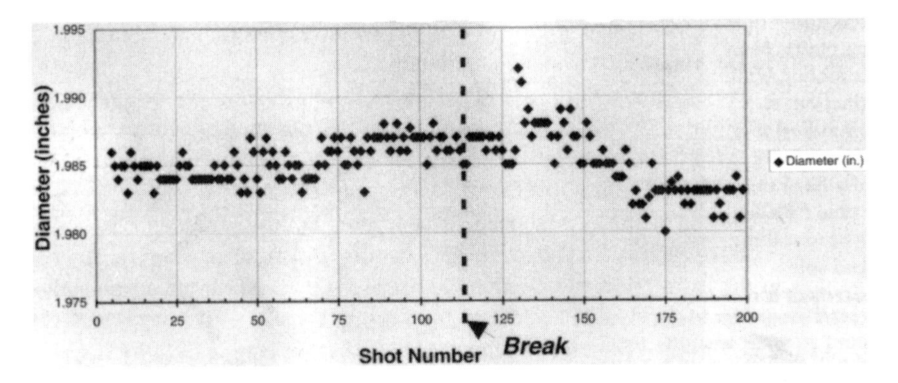


**Figure 9** Scatter diagram (dimension versus shot number) for 200 polystyrene parts injection molded in a Direct AIM tool.

The parts reported in Fig. 9 were injection molded on a 75-ton Van Dorn press at Ken McNabb Corp., Chatsworth, CA. The first 113 parts were injection molded on a Tuesday, with the remaining 87 parts being molded the next day. The break point is shown with a dashed vertical line. It is important to note that except for the two outlier points (numbers 130 and 131), 99% of the measurements are within ± 0.005 in. of the mean value, 95% are within ± 0.003 in., and 85% are within ± 0.002 in. While not yet at production tooling tolerances, these results are impressive for a technique that was first implemented as recently as 1995.

Note the cycle times for Direct AIM. Relative to production tooling, these cycle times are quite long. This is one of the reasons that Direct AIM is *not* appropriate for manufacturing large quantities of plastic parts. First, Direct AIM tools are useful for injection molding 20–50 parts. Beyond 50 cycles, the inserts, and especially the core, start to exhibit signs of wear. Second, 3–5min cycle times are *not* economical for large production lots.

Nonetheless, even at the slowest Direct AIM rate, a 5-min cycle will still enable 12 parts to be injection molded per hour, or fifty *parts* in *about* 4 h. How valuable would this capability have been on project Widget, where the team only needed 5 *parts* for thermal and mechanical testing? Although 5-min cycle times may seem agonizingly slow to an injection molder, if he realizes that the customer is actually buying *overall time saved*, spending less than 1 h injection molding five parts to help save 9 *weeks* is terrific.

Experience has shown that contrary to intuition, Direct AIM inserts are not primarily damaged during the *injection* process. Rather, they are far more commonly damaged during the *ejection* process. Apparently, longer cycle times allow the hot plastic inserts to cool below their glass transition temperature, thereby greatly increasing their strength and modulus. An effective procedure involves opening the press after the injection-molded plastic part has fully solidified, and then blowing cool air on the Direct AIM core *prior* to part ejection. The forced convection air cooling of the core will noticeably increase tool life. The use of any of a number of commercial release agents on every shot is also recommended. Again, although this does indeed add a few seconds to each cycle, the *overall* project time savings can be so considerable that the extra effort involving mold release is well worth the advantages related to mold survival and reduced time-to-market.

The primary advantage of Direct AIM is that the core and cavity inserts are generated *directly* on an SLA, with no secondary processes required other than preparation for installation on an injection-molding press. However, the shortcomings of Direct AIM are as follows:

1. The thermal conductivity of cured SL resin is about 300 *times lower* than that of conventional tool steels. As a result, the rate at which the tool can dissipate heat from the injected plastic is correspondingly diminished. The low thermal conductivity of SL resins accounts for the extended cycle times required when using Direct AIM inserts.
2. Large ACES inserts can involve 30–40 h build times on an SLA. At typical service bureau rates of $50–$70/h, this represents a significant cost.
3. The physical strength of Direct AIM inserts is poor, especially at the elevated temperatures encountered during injection molding. As noted earlier, tool damage often occurs during part *ejection*. The hot injected thermoplastic tends to stick to the ACES core. Additionally, the core has been softened and weakened as a result of its elevated temperature. Finally, as the plastic cools, it shrinks *onto* the core, making extraction even more difficult. Attempting to eject the molded part too soon can lead to core fracture.
4. Finally, the active surfaces of a Direct AIM insert are subject to damage through abrasion, as cured SL photopolymers are extremely soft relative to typical tool steels. Specifically, the injection of glass fiber-filled thermoplastics will substantially shorten the useful life of Direct Aim bridge tools.

To address these issues, a number of variations of the Direct AIM concept have been developed and tested by 3D Systems, as well as a growing list of users. In essence, these ideas involve different types of ''backing'' and ''fronting'' materials. The first of these variations is illustrated in Fig. 10.

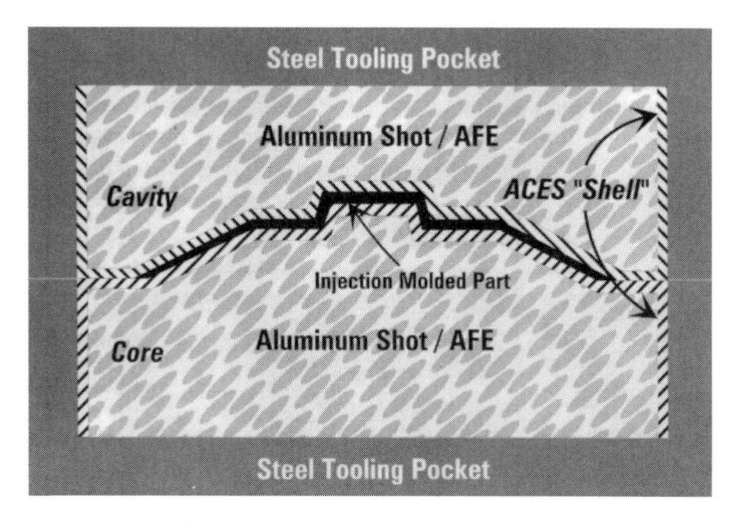


**Figure 10** Illustration of the concept of ''shelling'' a Direct AIM insert for backing with aluminum-filled epoxy.

As shown in Fig. 10, rather than building a ''solid'' ACES core and a ''solid'' ACES cavity, two relatively thin ''ACES shells'' are built on the SLA. The benefits are as follows:

1. Saves build time
2. Enables backfilling with aluminum-filled epoxy, which is considerably less expensive than SL resin
3. Provides enhanced thermal conductivity relative to fully cured SL photopolymer resin
4. Simplifies the implementation of conformal cooling through the use of bent copper tubing

Excessive shell thickness will reduce these benefits. Conversely, if the shell is made too thin, it can ''sag'' under gravity and will not retain dimensional accuracy. Test data obtained to data indicate that the best results are realized when the core and cavity *side walls* are about 2.5–3.0 mm thick, and the *active mold surfaces* have shells between 1.5 and 2.0 mm thick.

After the shells have been built and cleaned, their supports are removed, and the resulting parts are postcured. The shells are then turned upside down, and copper cooling lines are bent and positioned near the active surface, while conforming to the general shape of the final injection molded part. The remaining void space is finally backfilled with a mixture of aluminum powder and two-part epoxy resin.

In order to determine the effect of different ''backing'' materials on the overall thermal conductivity of the resulting core and cavity inserts, a series of tests were conducted. Standard test samples were prepared in the form of 50-mm-diameter, 10-mm-thick disks. Data were obtained for the following six cases:

1. ACES (SL 5170)
2. Composite aluminum-filled epoxy (CAFÉ)
3. CAFÉ with 20% by weight Al shot ($\approx$ 1–6 mm shot diameter)
4. CAFÉ with 40% by weight Al shot
5. CAFÉ with 60% by weight Al shot
6. An ACES 2-mm-thick shell, backed with CAFÉ, containing 40% by weight Al shot

Figure 11 shows the results of these tests. As a point of reference, the thermal conductivity values for copper, aluminum, and $A_6$ tool steel are also shown in the same units. Three important observations can be made from the data of Fig. 11:

1. The thermal conductivity values for the pure metals are 100 to 1000 times greater than those of the various composite bridge tooling materials.
2. CAFÉ with 60% aluminum shot has a thermal conductivity that is about an order of magnitude better than a straight ACES sample.
3. The composite–AIM sample (i.e., a 2-mm-thick ACES Direct AIM shell, backed with CAFÉ mixed with 40% Al shot) had a thermal conductivity *about three times that of an ACES sample made from solid SL 5170 resin.*

From these results, it is clear that backing with appropriate materials can improve the thermal conductivity of Direct AIM tools. The improved thermal conductivity correspondingly reduces cycle time to about 2 min, from the roughly of 4–5-min cycle times for *solid* direct AIM inserts. An extension of this concept by Morgan (11) was the use of low-melting-point alloys of bismuth, tin, antimony, and lead as backing materials for Direct AIM thin shells. Using the specific alloy CerroBend with a melting point of 58°C, ($\approx$136°F) to back a 1.5-mm-thick ACES shell built from SL 5170 resin, an effective thermal conductivity for the composite article was determined to be about 8 $\times 10^{-3}$ cal/s cm °C. This is over twenty *times better* than the thermal conduc-

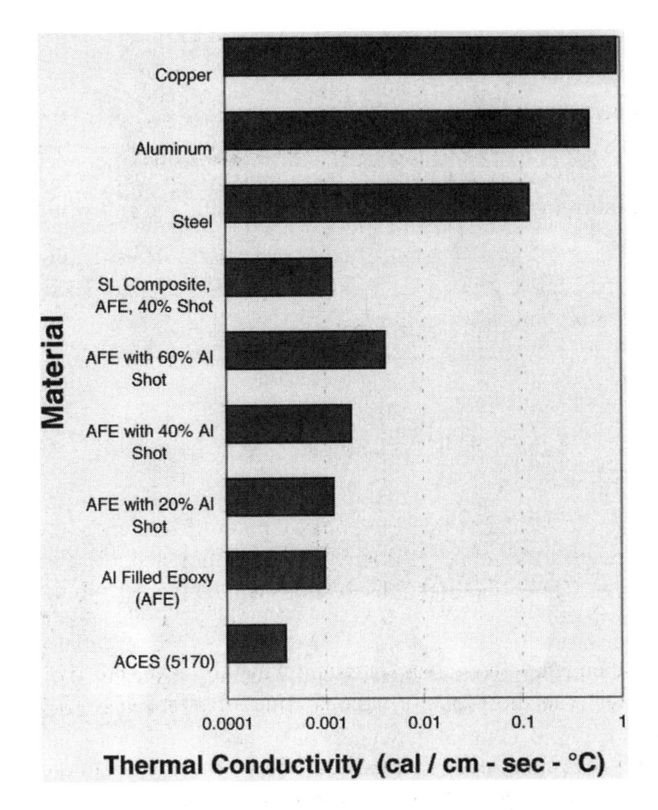


**Figure 11** Thermal conductivity values for some potential tooling materials.

tivity of a solid ACES insert of the same geometry, and almost 7 *times better thermal conductivity than an optimized composite–AIM insert.* Here, cycle times have been reduced to about 1 min.

An additional advantage of the low-melting-point alloy approach is that when a tool has completed its function, the backing material can be melted in boiling water and reused many times. With improved thermal conductivity for significantly faster cycle times, and enhanced compression strength for longer tool life, recycling the backing material becomes even more cost-effective.

In addition to *backing* Direct AIM shells, a number of organizations are currently studying various methods of *fronting* the active tool surface with a material, or materials capable of the following:

1. Improved heat dissipation
2. Higher surface hardness
3. Enhanced abrasion resistance

Work has been done utilizing electroless chemical deposition of copper on ACES inserts. In 1 case, 200 polycarbonate parts were successfully injection molded in a copper-coated Direct AIM tool (12). To further increase active surface hardness and abrasion resistance, other groups are investigating the use of electroless nickel deposited on top of electroless copper. Techniques, including

1. Physical vapor deposition
2. Chemical vapor deposition
3. Low-temperature spray metal deposition
4. Pulsed laser deposition

are all currently being evaluated.

An important aspect of this research involves the adhesion of the fronting material throughout hundreds to perhaps thousands of injection cycles. Unless the coefficient of thermal expansion of the fronting material closely matches that of the substrate or the adhesion between the two is excellent, the inevitable stresses resulting from expansion and contraction during each injection/cooling cycle may indeed cause the coating to spall off the substrate. This would significantly diminish the effectiveness of such a fronting material. Hopefully, combinations of materials, coating thickness, and process parameters can be found which will provide rugged, durable coatings.

## VI. RAPIDTOOL RAPID BRIDGE TOOLING

An altogether different approach to bridge tooling is the RapidTool™ process developed by DTM Corporation, Austin, TX. Here, the fundamental build technique is selective laser sintering (SLS). The SLS concept was originally developed by Carl Deckard, while a graduate student at the University of Texas, Austin, TX. The working materials initially involved various thermoplastics, including poly(vinyl chloride) (PVC), polycarbonate (PC), nylon, and wax powders, with the latter being used to fabricate patterns for shell investment casting.

DTM has extended the SLS process to include additional materials (13). These include (a) an acrylic-based powder called TrueForm PM, (b) a composite material consisting of nylon and glass-bead-reinforced nylon, called Proto-

Form, and, finally, (c) polymer binder precoated low-carbon-steel particles for RapidTool (14).

The RapidTool process is directly relevant to bridge tooling. It uses a 50-W $CO_2$ infrared laser emitting at 10.6 μm. The focused laser spot is scanned with a pair of orthogonal mirrors. The energy absorbed from the moving laser spot selectively fuses a thermoplastic polymer binder which has been precoated at a thickness of approximately 5 μm onto low-carbon-steel particles. The laser-fused binder holds the quasispherical steel particles together as a "green" part. In this state, the part is rather fragile, with a green strength of only about 440 psi, so care must be exercised to avoid damage to thin sections during handling. The quasispherical low-carbon-steel particles have a mean size of ≈55 μm, with a size distribution extending from about 30 μm to roughly 75 μm.

The green part is then placed in an electrical resistance furnace. Using a 25% hydrogen/75% nitrogen reducing atmosphere, the binder is almost totally eliminated when the furnace temperature reaches about 700°C (≈ 1300°F). The primary reduction product is methane ($CH_4$). Nitrogen and excess hydrogen will also exit the furnace. For environmental reasons, it is best if the $CH_4$ and any excess $H_2$ are passed through an afterburner, to enable combustion in the presence of abundant ambient air.

At elevated temperatures, the final combustion products will be primarily carbon dioxide and water vapor. Because any combustion process is never perfect, trace amounts of carbon monoxide and various oxides of nitrogen will also be generated. As they are produced in very small absolute quantities and can economically be mixed with large amounts of excess air, the final concentrations of CO and $NO_x$ can be made sufficiently small to satisfy even the most stringent environmental regulations.

Furthermore, just as any oxidation process is never perfect, neither is the reduction process used to eliminate the binder. Small amounts of carbonaceous residue will always remain and can actually act as a "glue" to temporarily help hold the steel particles together. The small passageways that result from the near elimination of the binder produce a porous article having about 60 vol% metal, and 40 vol% void space.

During the *single secondary furnace cycle* of the RapidTool process, (a) the polymer binder coating is eliminated, (b) the steel powder is sintered, and, finally, (c) the porous steel skeleton is infiltrated with copper. The infiltration is accomplished by placing solid copper slugs on top of the green part prior to the furnace cycle. When the furnace is heated to a temperature just above the melting point of pure copper (1083°C, or 1981°F), but well below the melting point of the low-carbon-steel particles, the molten copper then "wicks" into the part through capillary action. The result is an essentially fully dense part consisting of (a) the sintered low-carbon-steel particles and (b) their interstices which have subsequently been infiltrated with copper. The total linear shrinkage occurring in the furnace is approximately 2.5–3.5% (14).

The primary advantage of the RapidTool process is that it forms a *metal part* as its *direct* output, albeit after a secondary "binder elimination/steel particle sintering/copper infiltration" step. The resulting low-carbon-steel/copper part can be used as a core or cavity insert for rapid tooling. However, the elimination of the binder, the sintering operation, and the subsequent infiltration process involve significant linear shrinkage, as noted earlier. If this shrinkage was *absolutely constant* for all geometries, at say 2.5%, it would then be a simple matter to account for the entire process shrinkage by increasing the scale of all CAD dimensions by a factor of 1.025. Unfortunately, there are two fundamental problems with this approach.

First, the shrinkage process is almost *never* perfectly uniform. In investment casting, sand casting, die casting, plastic injection molding, as well as selective laser sintering, solid ground curing, fused deposition modeling, stereolithography, and, in fact, any process where there is a change of phase and an accompanying shrinkage, careful experimental measurements invariably show that thick sections will shrink somewhat differently than thin sections. Second, as we shall discuss in the following section, is the issue of "random-noise" shrinkage.

## VII. SHRINKAGE VARIATION

Many commercial processes involve a change of phase. Specifically, a material may be transformed from a liquid to a solid, or in some cases from a solid to a liquid and then back to a solid again. Some important examples are as follows:

1. Investment casting (solid metals are melted, poured into a ceramic mold, allowed to cool, solidify, and are removed from the mold)
2. Sand casting (similar phase sequence)
3. Die casting (similar phase sequence)
4. Injection molding (thermoplastics are melted, injected into a mold, cooled, solidified, and ejected)
5. Blow molding (similar sequence)

Also, within the field of RP&M, similar phase change phenomena occur. Specifically the following:

1. Stereolithography (liquid photopolymers are solidified by scanned ultraviolet laser radiation)
2. Solid ground curing (liquid photopolymers are solidified by flood ultraviolet radiation)
3. Selective laser sintering (thermoplastic powders or polymer coatings on metal powders are melted by infrared laser radiation, cooled, and solidified)
4. Fused deposition modeling (thermoplastics are melted, extruded, cooled, and solidified)

In each case, the phase change, from liquid to solid, involves a decrease in specific volume and a resulting shrinkage. The total volumetric shrinkage varies from process to process and from material to material, for a given process. However, *all* of these processes involve *some* level of volumetric shrinkage. Note that the volumetric shrinkage, $S_v$, and the linear shrinkage, $S$, are related, for the case of perfectly isotropic shrinkage, by the expression

$$S = 1 - (1 - S_v)^{1/3} \tag{1}$$

From the binomial theorem,

$$(1 + x)^n = 1 + \frac{nx}{1!} + \frac{n(n-1)x^2}{2!} + \frac{n(n-1)(n-2)x^3}{3!} + \cdots \tag{2}$$

We now set $x = -S_v$ and $n = 1/3$. For the case where $S_v \ll 1$, we can neglect higher-order terms involving $S_v^2$, $S_v^3$, and so on. Thus, $(1 - S_v)^{1/3} \approx 1 - 1/3\, S_v$. Substituting this result into Eq. 2, we obtain the often-used relationship

$$S \approx \frac{1}{3} S_v \tag{3}$$

Consequently, from Eq. (1), a material exhibiting 3% volumetric shrinkage should be undersize in all directions by $[1 - (1 - 0.03)^{1/3}] \approx 1 - 0.989898299 \cdots \approx 0.0101017 \cdots \approx 1.01017 \cdots \%$, provided the shrinkage was perfectly isotropic. Note that by neglecting higher-order terms, the simplified approximation of Eq. (3) would lead to a shrinkage of 1.00000% in all directions. Although Eq. (3) is certainly a close approximation, it is worth noting that the difference between the two results, (i.e., 0.01017%) is hardly as trivial as one might first suppose when attempting to generate highly accurate core and cavity inserts. For example, if we were working to develop a 20-in.-long insert, the error associated with the approximation of Eq. (3) would account for an error in the tool of 0.002034 in. all by itself. This is comparable to the entire error budget for production tooling!

Furthermore, it is important to remember that a fundamental assumption leading to this result was that the shrinkage is perfectly isotropic (i.e., identical in all directions). Unfortunately, this is rarely the case for real parts! At the atomic or molecular level, shrinkage may indeed be almost perfectly isotropic (15). However, because cooling will always occur preferentially at the surface, interior part temperatures will inevitably lag exterior part temperatures during the cooling process. Consequently, shrinkage will tend to occur initially at the outer perimeter of a part, and somewhat later within the interior of the part. The result, even for a simple thin-wall section, would be slightly different conditions acting on the central region of the part. This effect alone could account for tiny variations in overall part shrinkage. Additionally, constrained shrinkage associated with real part geometries (e.g., a thin-wall section joining a thick-wall section) will also result in numerous small shrinkage variations.

Notwithstanding these issues, the basic approach used in all of the processes noted earlier involves some form of shrinkage compensation. Traditionally, one experimentally measures the linear shrinkage for a given material in a given process, and then applies a ''shrinkage compensation factor'' to all part dimensions. The part is intentionally built oversize, so that when the inevitable process shrinkage occurs, the resulting part dimensions will be ''correct,'' if the calculations have been done properly. This sounds nice in principle. Unfortunately, experience indicates that *it is not a simple matter to achieve precise dimensional control through shrinkage compensation.*

## VIII. BACKGROUND

During the development of sterolithography, an important goal was improved part accuracy. A relevant story involves a series of events which occurred about 1990–1991. A potential customer had indicated that he would buy not only one but *two* SLA systems provided the customer's test part could be built such that 10 critical measurements each would fall within ± 0.005 in. of the respective CAD dimension.

A young and very enthusiastic applications engineer (AE), whose reputation shall be protected by anonymity, eagerly accepted the challenge. His plan was quite simple. Intentionally build the part with no shrinkage compensation whatsoever. Then, clean the part, remove the supports, postcure it, and very carefully measure all the resulting dimensions. Then, after the fact, determine the appropriate ''best-fit'' shrinkage compensation factor, rebuild the part accordingly, and ''voila'', all the dimensions should be ''right on.'' He did this. Unfortunately, it did not work. Many of the dimensions were well outside the allowable tolerance band of ± 0.005 in.

What could be wrong? Upon carefully reexamining the data, the young AE noticed that different dimensions on the part seemed to have slightly different shrinkage compensation factors. Unfortunately, the SLA software at that time could only accommodate a single shrinkage compensation factor. Later, the system software would allow different shrinkage compensation factors for *X*, *Y*, and *Z*, but even in 2000, one cannot use different values for each section thickness. What to do?

After some thought, the young AE came up with a very interesting idea. What if he went back to the original CAD model and modified each and every dimension according to its own experimentally determined, shrinkage compensation factor? Although tedious, if done properly, surely this would work. So he spent many hours painstakingly modifying the CAD model. When finished, he built the part a *second* time, cleaned it, removed the supports, postcured it, measured it, and, so forth; to his utter frustration, numerous dimensions were still outside the acceptable ± 0.005-in. tolerance band.

Convinced that his method would work, and hardly lacking in zeal, persistence, or motivation, he built the part a *third* time, repeating the original experiments all over again, while taking special pains to be extremely precise in his measurements. Indeed, the second set of shrinkage compensation values were slightly different than the original set. Surely, this must be the answer. He simply was not sufficiently careful the first time. Again, painstakingly modifying the CAD data, he built the part for the *fourth* time. After cleaning, support removal, postcure, and measurement, three dimensions were *still* outside the acceptable tolerance band. With a look of utter frustration mixed with resignation, the young AE finally abandoned the project, and shortly afterward left to accept another job elsewhere. Hopefully, his zeal and persistence have reaped more and better harvests.

What is the point of this story? The author should have understood the basic concept back in 1991. Unfortunately, insights do not always arrive like the cavalry in Westerns (i.e., just when you need them). In this case, it took about 7 years to put together the pieces of the puzzle. During that period, work with SL photopolymer shrinkage initiated the quest for a better understanding of shrinkage variation (16). Later, during the development of QuickCast, test results involving investment casting pointed to similar problems when trying to account for metal shrinkage (17). Still later, results involving 3D Keltool suggested that very similar phenomena were at work (18).

Furthermore, reviewing data from 3M Corporation on Tartan Tooling (19), data from the early ExpressTool (20), powder metallurgy process, as well as DTM Corporation's RapidTool (21), the results show a very familiar pattern. Also, in powder metallurgy, it has been known for some time that structural powder reorganization effects can lead to fluctuations in shrinkage (22). At a microscopic level, the spaces between individual particles are not identical and neither are the exact shape or orientation of neighboring particles. Thus, during sintering, the shrinkage will vary ever so slightly from one location to another, or from one run to the next. Parts made by the *same* process, using the *same* materials, with the *same* equipment run by the *same* trained people in the *same* environment are rarely ever identical.

What seemed evident to this author was not that the young AE had a bad idea or made sloppy measurements, neither are numerous investment-casting foundries lacking in skill, technique, or motivation. The same is surely true of the capable scientists and engineers at 3M, 3D, ExpressTool, and DTM. What gradually dawned as a possible explanation for the inability to precisely apply shrinkage compensation in real parts, was the concept of *random-noise shrinkage*.

## IX. RANDOM-NOISE SHRINKAGE

From experience we know that the shrinkage process is almost never perfectly uniform. In investment casting, sand casting, die casting, plastic injection molding, as well as SL, SLS, SGC, and FDM, and, in fact,any process where there is a change of phase and an accompanying shrinkage, careful experimental measurements invariably show that thick sections will shrink differently than thin sections.

Assume that we build a test part $N$ times using (a) the *same* hardware, (b) the *same* procedure, and (c) the *same* parameters, while holding (d) the environmental conditions *as constant as possible*. Measuring the dimension of each section and comparing this measurement with the intended CAD value for that dimension, we can define the linear shrinkage, $S_{j,i}$ for the $i$th measurement of the $j$th section, by the relation

$$S_{j,i} \equiv \frac{L_{j,\mathrm{CAD}} - L_{j,i}}{L_{j,\mathrm{CAD}}} \tag{4}$$

where $L_{j,i}$ is the $i$th measurement of the length (or width or height) of the $j$th section and, $L_{j,\mathrm{CAD}}$ is the *intended* CAD length (or width or height) of the $j$th section.

Note that shrinkage is dimensionless, as it involves a length divided by a length. We now define the mean shrinkage for the $j$th section, $\bar{S}_j$, in the usual manner, by summing the $N$ separate shrinkage values, and then dividing by $N$. In mathematical notation,where a bar over a quantity indicates the mean value of that quantity,

$$\bar{S}_j = \frac{1}{N}\sum_{i=1}^{N} S_{j,i} \tag{5}$$

Results for measurements on a single section thickness from $N = 30$ different SLA shrinkage test parts (16) are illustrated in Fig. 12. Repeating this procedure for each of the six different section thickness values (2.5, 3.75, 5.0, 7.5, 10.0, and 12.5 mm), the results are shown in Fig. 13. Four trends are evident from the data of these two plots:

1. There is a small but definite *variance* in the individual shrinkage values, $S_{j,i}$, for a *single dimension*, even though the same hardware, software, build procedure, and materials were used to generate the parts. This shows up as scatter in the data and is indicated by the presence of error bars.

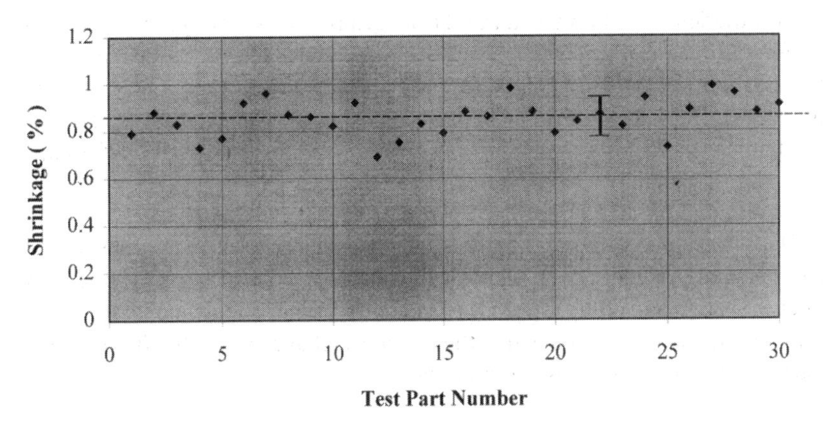


**Figure 12** CMM measurements on a single section thickness for 30 stereolithography shrinkage test parts.

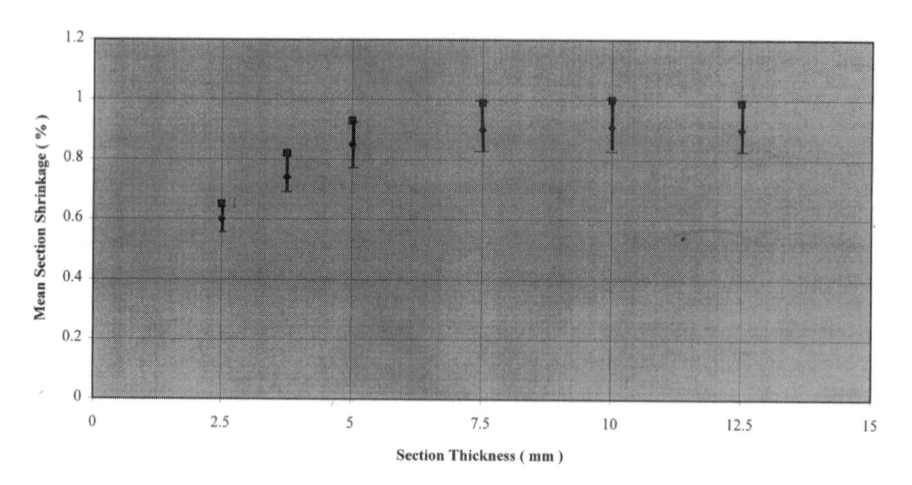


**Figure 13** CMM shrinkage measurements for six different sections.

2. There is also a small but definite *variance* between the mean shrinkage, $\overline{S}_j$, for each of the six different section dimensions.
3. Differences between the individual shrinkage values, $S_{j,i}$, and the mean shrinkage value, $\overline{S}_j$, for a given section are *not* trivial. The desired, albeit not easily achieved tolerance for production rapid tooling, is ± 0.002 in. on a 20-in. dimension. This corresponds to an error of ± 1 part in 10,000 or ± 0.01%. To assure 99.7% of dimensions will remain within this tolerance, three times the shrinkage standard deviation must not exceed ± 0.01%. Consequently, the standard deviation of the shrinkage, $\sigma_s$, should be less than 0.0033%, or about 1 part in 30,000. Tiny variations unimportant in generating a rapid prototype for concept validation become critical when attempting rapid tooling!
4. Finally, the difference between the *mean shrinkage* for one section thickness, $\overline{S}_j$, and that for another section is also nontrivial at the same level as that discussed in observation 3.

Two fundamental conclusions follow from Figs. 12 and 13 and the four observations listed. The first, embodied in observations 1 and 3, is that nontrivial differences in shrinkage occurring between otherwise *identical* sections, of otherwise *identical* parts, prepared by *identical* people, in an *identical* manner, on *identical* equipment, using *identical* materials, under nearly *identical* environments is a classic example of *random noise*!

The italics on the word *identical* are intended to remind the reader that no two parts and no two experiments, and indeed no two measurements within a given experiment are ever truly identical. In RP&M, tiny changes in temperature, pressure, humidity, laser power, and laser spot size will all effect the outcome. With metal powders, the binder composition, particle size distribution, particle shape, part handling, and the details of the mixing procedure have a finite influence.

In short, it appears there will always be a component of ''random-noise shrinkage'' superimposed on the mean process shrinkage. For convenience of expression, let us define the mean value of the shrinkage for a given process by $\bar{S}$ and the standard deviation of that shrinkage by $\sigma_s$. Note that because shrinkage is itself dimensionless, then $\sigma_s$ is also dimensionless. We will return to the matter of random-noise shrinkage shortly.

The second important conclusion from observations 2 and 4 involves differences between the mean shrinkage, $\bar{S}$, for different sections. The data imply that shrinkage is fundamentally *nonuniform.* Indeed, the shrinkage is dependent on the thickness of any section. Although the dependence is weak, it is finite. Again, although a single shrinkage compensation factor is good enough for prototype visualization, verification, and perhaps even iteration, it is *not* sufficient when generating SL patterns for production rapid tooling, or when using sintered powder metallurgy techniques for production core and cavity inserts. The use of separate shrinkage compensation factors for dimensions in the *X*, *Y*, and *Z* coordinate directions is an improvement, but even this approach does not account for variations in section thickness along the same coordinate.

To achieve the accuracy levels required for production rapid tooling, without postmachining, a more comprehensive method is required. The most successful approach to date has been employed by some investment casting foundries (23). It involves developing a body of experimental data for the measured shrinkage values of a great many different shapes and then applying slightly different shrinkage compensation factors to the CAD design for each section of a part. The major advantage of this approach is improved part accuracy. The disadvantages are as follows:

1. Considerable testing and experience are needed to establish a library of shrinkage compensation factors for an extensive repertoire of part shapes.
2. Applying multiple shrinkage factors is tedious, especially for complex geometries.
3. The Law of Universal Perversity virtually guarantees that as soon

as one thinks that the shrinkage data library is complete, a geometry will immediately be encountered for which no data exist!

However, although geometry dependent shrinkage is a nuisance, random-noise shrinkage is actually worse, precisely because it is both fundamental and random! It is one thing to spend the time to carefully gather a library of shrinkage compensation factors on the expectation that this will substantially improve the accuracy of the final parts. It is quite another to realize that *the random component of the shrinkage may already be greater than the allowable rapid tooling tolerance*!

In this type of situation, there is absolutely no way to predict what the exact value of the shrinkage will be for any section, simply because the random component of the shrinkage is indeed *random*. Here, Gaussian statistics will determine if the core or cavity in question will satisfy the tolerance specifications. In short, building a truly accurate pattern or tooling insert based on the use of a ''best value'' shrinkage compensation factor becomes a rather expensive roll of the dice. Occasionally, the random noise will be very near zero, the shrinkage compensation procedure will ''work,''and the result will be an accurate pattern or tooling insert. Unfortunately, the random noise will often not be trivial and the resulting pattern or tooling insert will not meet specification.

At this point, the reader may feel that this reasoning automatically implies that approaches based on the use of RP&M patterns and powdered metallurgy are doomed to failure as a means of generating accurate, reliable, and consistent core and cavity inserts. However, it is very important to note that the key to accuracy for such methods hinges on whether the random component of the shrinkage is greater than or smaller than the allowable rapid tooling tolerance. If we assume that the random component of the shrinkage obeys Gaussian statistics, as indeed most random phenomena do, and if we also define the acceptable rapid tooling tolerance as $\pm\epsilon_T$, for a dimension of length $L$, a reasonable criterion for the allowable level of random noise shrinkage can be written as

$$3\sigma_s L \leq |\epsilon_T| \qquad (6)$$

This concise relation assures that provided three times the standard deviation of the shrinkage times the length of the relevant dimension is less than or equal to the absolute value of the rapid tooling tolerance, then more than 99.7% of all such dimensions should lie within that tolerance.

If this were the case, the process would at least be *capable* of reliably providing accurate core and cavity inserts. If the mechanical properties, abrasion resistance, thermal conductivity, lifetime, cost, tool-generation lead time, and production cycle time were all either adequate or distinctly advantageous, rapid tooling by such methods could and almost certainly would become a practical reality.

Unfortunately, as is often the case with many ''terrific ideas,'' a major barrier to progress can reside in the numbers. Although many believe that $\pm$ 0.002-in. ($\approx$50-μm) tolerances are only required on critical dimensions, and especially at parting surfaces and shutoffs, the point is they *are* required at certain locations. Either we accept this part of the rapid tooling challenge or postmachining will always be required.

Let us therefore assume as a most stringent case that $\epsilon_T = \pm\ 0.002$ in. The left side of relation (6) is smaller than the right side until $L$ reaches its maximum allowable value, $L_{max}$, at which point the equal sign applies. When this condition occurs, the allowable standard deviation of the shrinkage is given by

$$\sigma_s = \frac{|\epsilon_T|}{3L_{max}} \tag{7}$$

If we assume that $L_{max} = 20$ in. would suffice for the great majority of rapid tooling applications, then the allowable standard deviation of the process shrinkage would be

$$\sigma_s = \frac{0.002 \text{ in.}}{3 \times 20 \text{ in.}} = 0.000033 = 0.0033\%$$

Think about the implications of this result. If the mean shrinkage for a process is quoted as 0.8%, but the actual value is really 0.79%, this would hardly seem like a problem, right? WRONG! The difference between 0.80% and 0.79% is obviously 0.01%. In the spirit of clarification, some simple numerical examples are as follows:

1. 1% of 20 in. is 0.200 in.; grossly outside production tolerance.
2. 0.1% of 20 in. is 0.020 in.; well outside production tooling tolerance.
3. 0.01% of 20 in. is 0.002 in., which appears to just meet production tooling tolerance. However, this is only the value for one standard deviation! Whereas approximately 68% of the part measurements would be within production tooling tolerance, unfortunately, about 32% of the measurements would still fall *outside* that tolerance.
4. 0.0033% of 20 in. is 0.00066 in. This implies that 99.7% of all part measurements will indeed be within tolerance. This is the level of

> shrinkage uncertainty required for production rapid tooling. Unfortunately, the shrinkage for a given process would have to be quoted, (e.g., as 0.037 ± 0.003%). To the best of the author's knowledge, *none* of the existing powder-metallurgy, based rapid tooling processes even begin to specify shrinkage at this low level, or with this kind of statistical precision.

3D Keltool, RapidTool, Phast (24), and ExpressTool's earlier powder metal technique, all utilize some form of powder-metallurgy process. Each involves a phase change and consequent shrinkage. How then could techniques such as these ever hope to produce core and cavity inserts of sufficient accuracy to enable production rapid tooling without subsequent machining? The answer lies in reducing the value of $\sigma_s$. Of course, this sounds logical, but how does one actually do this?

Since 1989, this author, as well as many co-workers, have been acutely aware of, and deeply involved with the effects of shrinkage on the accuracy of SL parts. Further, with the advent of the QuickCast process, similar problems involving the effects of shrinkage on the dimensional accuracy of investment cast parts also became evident. Additional studies, as well as discussions with experts from various other metal-forming and plastic-injection-molding disciplines indicated the existence of related problems, albeit at different levels of shrinkage and distortion. Shrinkage-related errors also occur in sand casting, die casting, injection molding, as well as all the RP&M techniques.

## X. RANDOM-NOISE SHRINKAGE HYPOTHESIS

At this time, a limited amount of precise, statistically significant shrinkage variation data exists. Some data were compiled by 3M Corp. (25,26) during their invention, development, test, and commercialization of the Tartan Tooling process from about 1972 to 1986. In 1987, this process was sold to Wayne Duescher of St. Paul, MN and was renamed the Keltool process. The procedures and related intellectual property were again sold in 1996, this time to 3D Systems, Inc., and were subsequently renamed the 3D Keltool™ process.

Figure 14 plots the linear shrinkage for a single dimension, from 30 otherwise identical test parts, for one specific combination of sintering materials and process parameters. Note the variance in the measured values of the shrinkage from test part to test part. For this case, the mean process shrinkage is $\bar{S} = 0.799\%$ and the standard deviation of the shrinkage is $\sigma_s = 0.077\%$. Taking the ratio $\sigma_s/\bar{S}$, we obtain 0.096.

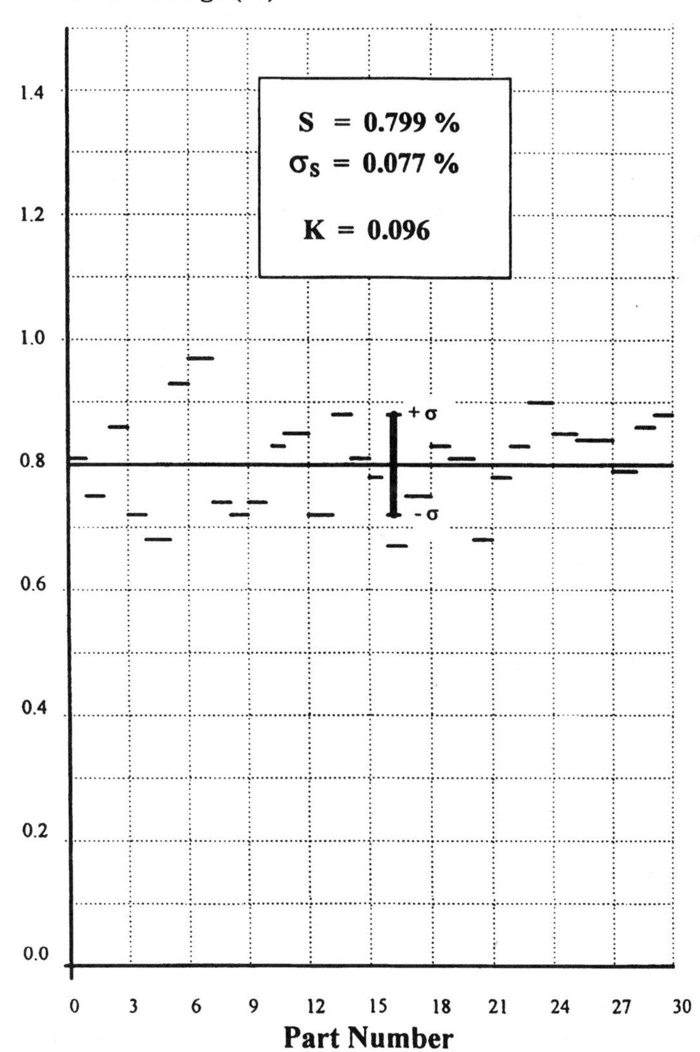


**Figure 14** Linear shrinkage versus part number for the large-shrinkage case.

Figure 15 plots similar data for a different set of sintering materials and process parameters. This time the mean process shrinkage is $\bar{S} = 0.402\%$ and the standard deviation of the shrinkage is $\sigma_s = 0.038\%$. Note that not only is the mean process shrinkage smaller for this case, but the standard deviation is also smaller. Again, taking the ratio $\sigma_s/\bar{S}$, we obtain the value 0.095, which is very close to the previous result.

Finally, Fig. 16 plots similar data for yet another set of sintering materials and process parameters. This time the mean process shrinkage is $\bar{S} = 0.201\%$ and the standard deviation of the shrinkage is $\sigma_s = 0.019\%$. Again, taking the ratio $\sigma_s/\bar{S}$, we obtain the value, 0.095, which is also very close to the results for the two other cases.

These data will support a hypothesis regarding shrinkage variation. Additional data are required for confirmation. Based on information available to the author and the results presented in Figs. 14–16, the following hypothesis is proposed:

For processes involving a phase change, the resulting random-noise shrinkage is directly proportional to the mean process shrinkage.

Mathematically, this statement takes the simple form

$$\sigma_s = K\bar{S} \tag{8}$$

where $K$ is a proportionality constant referred to hereafter as the random shrinkage coefficient for a given process.

The value of $K$ can be determined from statistically significant data for a given process. For the cases presented, $K \approx 0.096$. Equation (8) also implies the following:

1. All shrinkage phenomena involve both a mean process shrinkage, $\bar{S}$, as well as a *superimposed random-noise shrinkage* having a standard deviation $\sigma_s$.
2. The larger the mean process shrinkage, the greater its standard deviation.
3. Because the random noise shrinkage is indeed *random* and cannot be predicted or compensated in advance, the key to accuracy and repeatability for such techniques is the reduction of the mean process shrinkage, $\bar{S}$, to the smallest possible level.

As noted earlier, when $L = L_{max}$, the equal sign applies in relation (6). Substituting for $\sigma_s$, from equation (8), we obtain the important result

$$\bar{S}L_{max} = \frac{|\epsilon_T|}{3K} \tag{9}$$

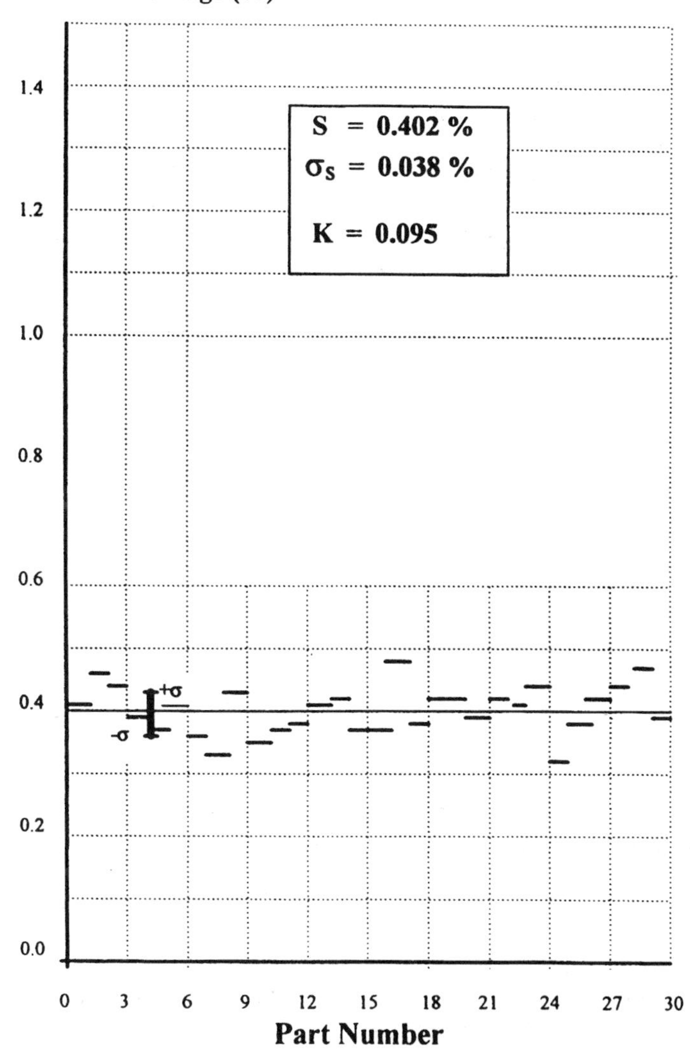


**Figure 15** Linear shrinkage versus part number for the intermediate-shrinkage case.

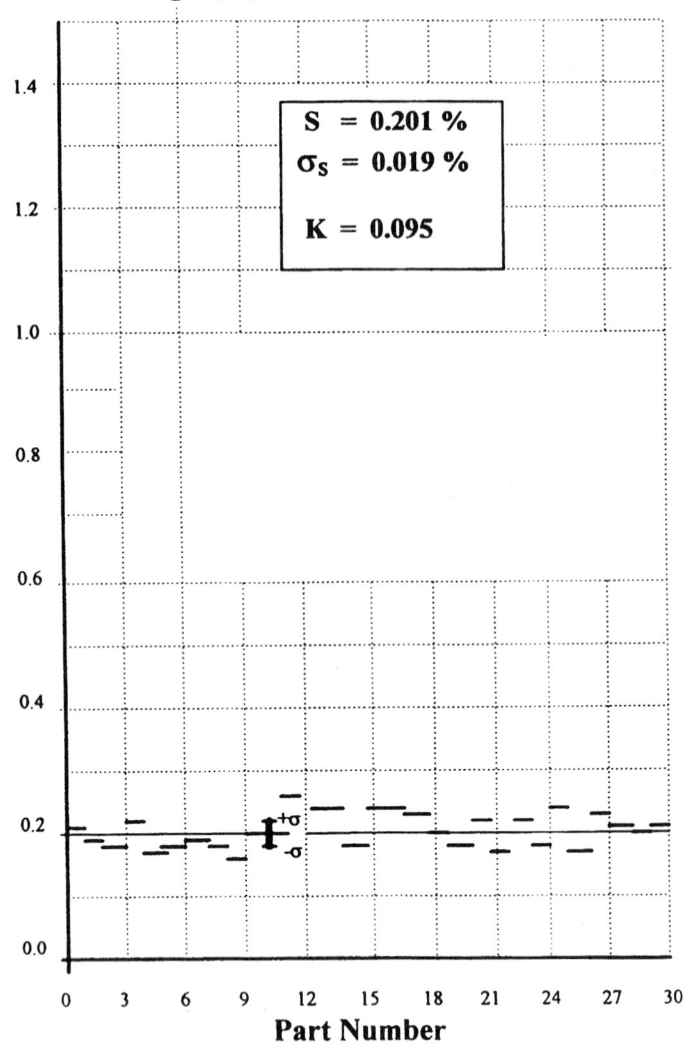


**Figure 16** Linear shrinkage versus part number for the low-shrinkage case.

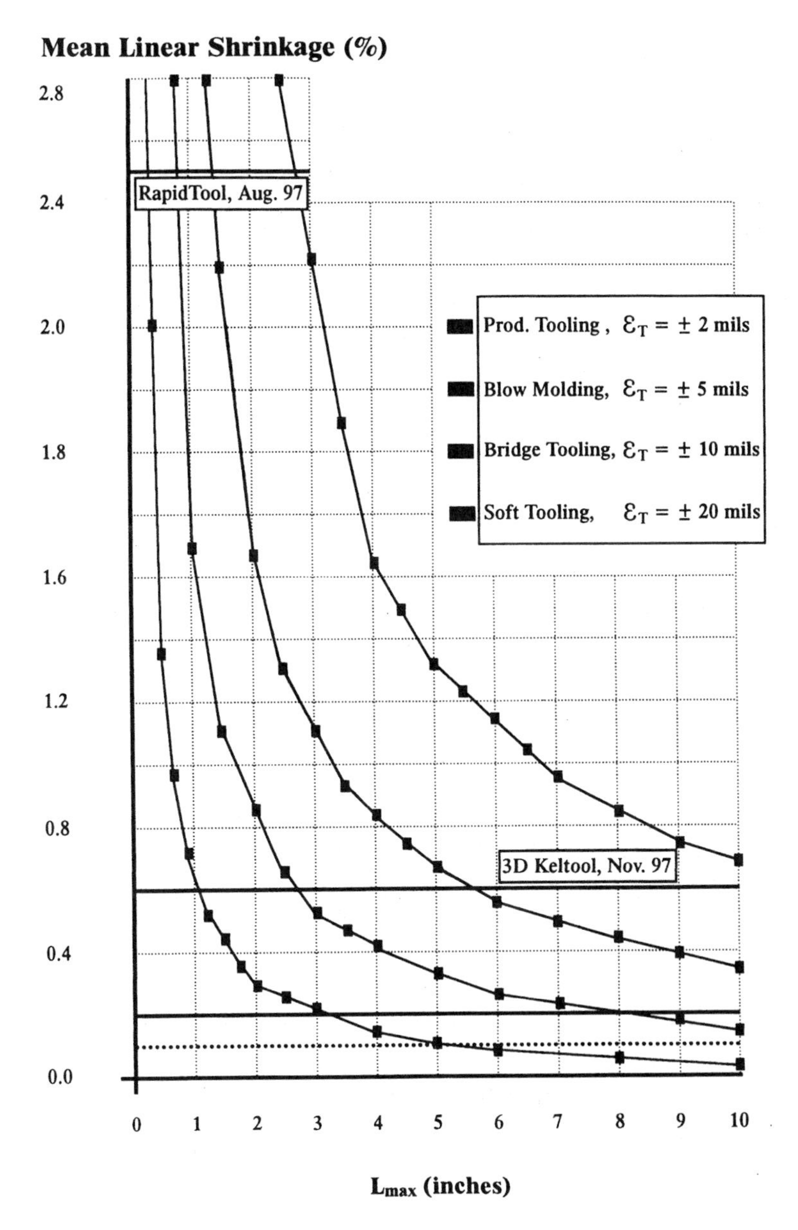


**Figure 17** Mean linear shrinkage versus length for various values of rapid tooling tolerance.

For a given process and tolerance, the right side of Eq. (9) is a constant. Thus, the mean linear process shrinkage, $\bar{S}$, and the maximum tooling insert length, $L_{max}$, capable of satisfying the required tolerance, $\epsilon_T$, are related *hyperbolically*! This is shown in Fig. 17 for the following four values of the rapid tooling tolerance, $\epsilon_T$: ± 0.002 in. for production injection-mold tooling, ± 0.005 in. for production blow-mold tooling, ± 0.010 in. for bridge tooling, and ± 0.020 in. for soft tooling.

It is quite clear from Figure 17 that until the mean process shrinkage is reduced to values less than 0.1%, the resulting random-noise shrinkage will make it very difficult to achieve 5-in. production rapid tooling dimensions on a consistent basis. Higher mean process shrinkage levels typical of current powder metallurgy processes allow some dimensions to satisfy the respective tolerances, but, unfortunately, others will not. Consequently, reliable rapid tooling inserts capable of achieving production injection-molding tolerances can presently be generated only for relatively small inserts.

Three potential solutions involve the following:

1. Research to reduce mean process shrinkage
2. Assuring all dimensions are "metal-safe" for final CNC machining
3. Using a process which involves zero or near-zero mean process shrinkage

Three such approaches are described in this book. One Involves nickel-vapor deposition and two involve electroforming (CEMCOM and ExpressTool). Because electroforming involves essentially zero mean process shrinkage, the problems associated with random-noise shrinkage are avoided. Furthermore, the ExpressTool process has demonstrated 20–30% cycle time reductions relative to CNC-generated tooling through the use of high-conductivity backing materials and conformal cooling.

## REFERENCES

1. C Hull, P Jacobs. Introduction to RP&M. In: P Jacobs, ed. *Rapid Phototyping & Manufacturing: Fundamentals of Stereolithography*. Dearborn, MI: SME Press; 1992, pp. 4–11.
2. L Andre, L Daniels, S Kennerknecht, B Sarkis. QuickCast™ foundry experience. In: P Jacobs, ed. Stereolithography and Other RP&M Technologies. Dearborn, MI: SME Press; 1996, pp. 209–237.
3. D Smock. New moldmaking systems slice art-to-part cycles. Plast World Mag 38–42, July 1995.

4. T Wohlers. Ten inventions that have forever changed product development. Seventh International Conference on Rapid Phototyping, San Francisco, 1997.
5. T Kerschensteiner. AMP Inc., A Simultaneous Engineering Case Study. In: P Jacobs, ed. Rapid Prototyping & Manufacturing: Fundamentals of Stereolithography. Dearborn, MI: SME Press, 1992, pp. 371–380.
6. P Smith, D Reinertsen. Achieving overlapping activities. In: Developing Products in Half the Time. New York: Van Nostrand, 1991, pp. 153–167.
7. S Willis. Real materials: Fast rapid tooling for injection molds. Proceedings of the Seventh International Conference on Rapid Prototyping, San Francisco, 1997, pp. 1–12.
8. E Gargiulo. In-plane stereolithography part accuracy. Proceedings of the 1st European Conference on Rapid Prototyping & Manufacturing, Nottingham, UK, 1992.
9. S Rahmati, P Dickens. SL Injection Mold Tooling. Surrey, UK: Prototyping Technology International, UK International Press, 1997.
10. B Bedal, H Nguyen. Advances in part accuracy. In: P Jacobs, ed. Stereolithography and Other RP&M Technologies. Dearborn, MI: SME Press, 1996, pp. 156–164.
11. W Morgan. Low melting point alloys as backing materials for Direct AIM™ plastic injection tooling. North American Stereolithography Users Group Meeting, Orlando, FL, 1997.
12. M Wilson, M Yeung. J Rapid Prototyping Tech 2(1), 1996.
13. K McAlea, P Forderhase, U Hejmadi, C Nelson. Materials and applications for the SLS® selective laser sintering process. Proceedings of the Seventh International Conference on Rapid Prototyping, San Francisco, 1997, pp. 23–33.
14. U Hejmadi, K McAlea. Selective laser sintering of metal molds: The Rapid Tool process. Proceedings of the Solid Freeform Fabrication Symposium, Austin, TX, 1996, pp. 97–104.
15. T Mueller, Plynetics Express, personal communication, 1998.
16. H Nguyen, J Richter, P Jacobs, In: P Jacobs, ed. Rapid Prototyping & Manufacturing: Fundamentals of Stereolithography. Dearborn, MI: SME Press, 1992, pp. 250–254.
17. L Andre, L Daniels, S Kennerknecht, B Sarkis. In: P Jacobs, ed. Stereolithography and other RP&M Technologies. Dearborn, MI: SME Press, 1996.
18. R. Connelly. Rapid tooling for medical products using 3D Keltool™. Proceedings of the Rapid Prototyping and Manufacturing '97 Conference, Dearborn, MI, 1997, pp. 89–99.
19. K Dillon, R Terchek. U.S. Patent 4,431,449, Feb. 14, 1984, (assigned to 3M Corp).
20. D Glynn, P Jacobs. CMM Measurements of Patterns and Powder Metal Inserts. ExpressTool internal report, January 1998.
21. K McAlea, P Forderhase, U Hejmadi, C Nelson. Materials and Applications for the SLS Process. Proceedings of the Seventh International Conference on Rapid Prototyping, San Francisco, CA, 1997.

22. W Soppe, J Janssen, B Bonekamp, L Correia, H Veringa. A computer simulation method for sintering in three dimensional powder compacts. J Mater Sci 29:754–761, 1994.
23. L Andre, Solidiform Inc., personal communication, 1997.
24. J Tobin. U.S. Patent 5,507,336, April 16, 1996.
25. K Dillon, R Gardner. U.S. Patent 4,327,156, April 27, 1982 (assigned to 3M Corp).
26. K Dillon, R Terchek. U.S. Patent 4,455,354, June 19, 1984, (assigned to 3M Corp).

# 5
# Rapid Production Tooling

**Paul F. Jacobs**
*Laser Fare—Advanced Technology Group*
*Warwick, Rhode Island*

**Larry André, Sr.**
*Solidiform, Inc.*
*Fort Worth, Texas*

## I. INTRODUCTION

We have discussed some of the ways that the Project Widget team could have utilized either *rapid soft tooling* or *rapid bridge tooling* to significantly reduce time to market. Through concurrent engineering practices they were able to save about 6.5 weeks; cutting a 63.5-week product development cycle down to 57 weeks. Furthermore, had they employed some of the cultural changes described in chapter 3, section I, they might have saved another 6 weeks, thereby slicing their product development cycle down to about 51 weeks. Nonetheless, even with these admirable efforts, Acme still would have beaten them to the market by about 2 weeks.

However, had the Widget team simply used *CAFÉ rapid bridge tooling*, they could have cut another 11.5 weeks off the product-development cycle, shrinking the latter from 51 weeks down to just under 40 weeks. Note that 40 weeks represents about a 37% reduction in the product-development cycle relative to 63.5 weeks *without* concurrent engineering, and almost a 30% reduction relative to a 57-week cycle *with* concurrent engineering! The use of CAFÉ *alone* would have enabled them to beat Acme to the marketplace by 2 months! Further, they would have saved about $40,000. Clearly, the gains associated with rapid bridge tooling are already quite significant and account for its accelerated utilization by industry during the past few years.

Notwithstanding all these benefits, the really *dramatic* advance will occur with the widespread implementation of *rapid production tooling*, which has been the long-term goal of both authors, as well as many others, since the early days of stereolithography. Surely, whatever size the worldwide *prototyping* market may be (reliable estimates have proved curiously elusive), it is clear that the equivalent market for tooling, as well as subsequent *manufacturing* of various components through injection molding, blow molding, die casting, powder injection molding, and investment casting, is likely at least two orders of magnitude larger.

Even if the techniques of rapid prototyping (RP) advance significantly in terms of accuracy, surface quality, materials, speed, and cost reduction, the RP market could be expected to encounter the first signs of saturation at revenues of about $700 million per year, with significant slowing of market growth at about $1 billion per year. Why then do so many people at numerous organizations continue to be bullish on the future growth of the rapid prototyping and manufacturing (RP&M) industry? The answer lies in the ''M.''

The world market for manufacturing the items noted above is so enormous that even if rapid tooling is only able to garner a small slice, the overall market for RP&M could increase by more than an order of magnitude over the next decade. Will this actually happen? The answer is probably yes. When something is needed by many and when that thing can save considerable time and money, the pressure to invent, develop, improve, and commercialize a practical, working version becomes very great. At least 25 different groups are currently investigating rapid production tooling. Will they all be successful? Probably not. Will one or a few of them be successful? Probably! In the words of the late sportswriter Grantland Rice, ''The race is not always to the swiftest, nor the battle always to the strongest, but that's the way to bet!''

## II. THE 3D KELTOOL PROCESS

As noted in Chapter 4, the 3D Keltool process is based on work performed from about 1972 until 1986 at Minnesota Mining and Manufacturing Corporation, now simply referred to as 3M Corp. The process, originally referred to as 'Tartan Tooling' by 3M, is fundamentally a powder metallurgy approach. It is shown schematically in Fig. 1.

The process starts with a master pattern. This is commonly in the form of a *positive* (i.e., identical to the final part geometry, except increased in scale

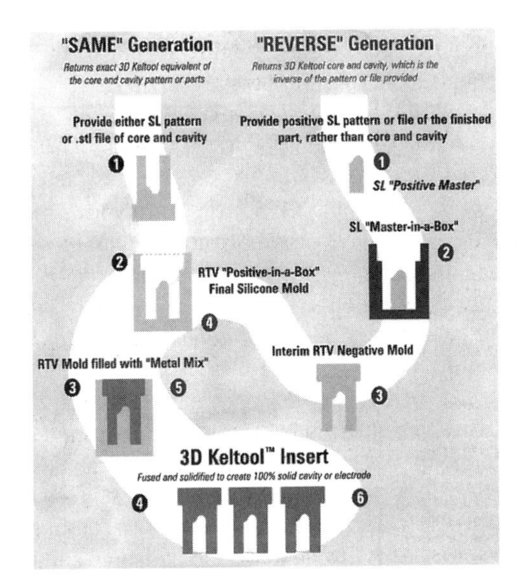


**Figure 1** A schematic flow diagram of the 3D Keltool process. (Courtesy 3D Systems.)

to allow for (a) any shrinkage involved in making the pattern, (b) the shrinkage of the 3D Keltool process, and (c) the shrinkage of the final injection-molded plastic). Alternatively, the pattern may be in the form of a *negative* (i.e., shaped like the core or cavity which will ultimately mold the final part, but again oversized to allow for the shrinkage values noted above). The pattern may be produced in various ways, including machining, grinding, carving, or by one of the various RP&M methods.

In fact, it was precisely the difficulty in rapidly achieving an *accurate pattern* that was a major stumbling block for 3M in the 1970s and the early to mid-1980s. It is ironic that during this period, 3M had developed a technique to rapidly generate *production tooling inserts and electrical discharge machining (EDM) electrodes*, but did *not* have access to a rapid source of accurate, high-quality, repeatable patterns. Consequently, much of the time saved generating the inserts or electrodes was offset by the lead time required to obtain accurate patterns. Coupled with some of the practical limitations of the Tartan Tooling process itself and the fact that neither the tooling cycle time reductions nor the cost savings were sufficiently great to establish a robust rapid tooling market at that time, 3M decided to sell the process in 1986. It is perhaps doubly ironic that about the time 3M decided to sell, 3D Systems was being incorporated to develop stereolithography, a process that would ultimately provide a rapid source of reasonably accurate and repeatable patterns!

As seen in Fig. 1, there are two different routes to a final 3D Keltool insert. One path is known as *reverse generation* because it returns to the customer a core or a cavity insert which is the *reverse* of the pattern or file provided. This procedure is shown on the right side of Fig. 1. The second approach, is known as *same generation,* because it returns a 3D Keltool core or cavity insert from a pattern or file shaped the *same* as the core or cavity (except for the process shrinkage noted previously). This route is shown on the left side of Fig. 1. The reader will note that the *reverse generation* method requires six distinct steps to achieve a tooling insert, whereas the *same generation* technique involves only four separate steps.

Because additional steps always introduce entropy, any transfer process is never perfect. Why then would anyone intentionally pick the process that involves two extra steps? The answer is because it is much easier to sand and polish a positive pattern than a negative pattern. Although this may seem like a small point, it is not. Experience has shown that sanding and polishing complex *negative* pattern geometries, to remove stair-stepping artifacts, can be very time-consuming. It is not unusual to hear stories of a negative stereolithographic (SL) tooling pattern taking 1.5 days to build on an SLA, but then needing 4–5 days just to sand and polish!

Remember, we are trying to reduce the *overall* time-to-market. Spending 5 days sanding and polishing a pattern, to save two additional 3D Keltool process steps that require a total of only about 2 days is not the way to win at this game. An interesting strategy used by some customers is to send a same generation pattern of the core (which is intrinsically a positive and hence relatively easy to polish), in order to save Keltool *process* time, and a reverse generation pattern of the cavity, which is therefore also a positive, to save *finishing* time.

The next step involves creating an room-temperature vulcanized (RTV) silicone rubber "Positive-in-a-Box" intermediate mold. As seen in Fig. 1, this may involve either one step or three steps. As discussed in Chapter 4, RTV molds can faithfully reproduce fine detail. Also, their high flexibility enables the removal of fragile green compacts with reasonable yield. Unfortunately, as also noted earlier, the CTE of typical RTV silicone rubber compounds is $\sim 300 \times 10^{-6}$ mm/mm °C. This is roughly 20 *times* the value of typical tool steels! For a mold having a maximum linear dimension of 500 mm (~20 in.), a temperature difference in the RTV mold of only 1°C (1.8°F) will result in its linear expansion by 0.15 mm, or 150 μm, or about 0.006 in.!

The reader will quickly recognize that this source of error *alone* is three times the desired rapid-tooling tolerance for plastic injection molding. Furthermore, this example involved a temperature difference of only 1°C! Consider the effects of the inherent silicone rubber exotherm, or what happens when somebody opens a door to the process room on a hot day.

The good news about making intermediate molds from RTV is that the surface replication is excellent, and the flexibility of RTV assists in the demolding process. The bad news is that without precise and costly process temperature control, the intermediate mold may be larger than, smaller than, or fortuitously equal to the size of the pattern. Unfortunately, these variances may *not* be small relative to production rapid-tooling tolerances.

After the RTV 'Positive-in-a-Box' intermediate mold has completely cured (hopefully at the correct temperature), a special bimodal mix consisting of (a) $A_6$ tool steel particles and (b) tungsten carbide (WC) particles is blended with (c) a proprietary binder. The blending process is performed with a high-torque, water-cooled sigma mixer. Although thorough blending is certainly important for achieving a uniform consistency, the blending cannot continue for more that about 10 min or the binder will begin to cure. This would greatly increase the viscosity of the mixture and impede complete RTV mold filling.

The 3D Keltool ''bimodal'' particle size distribution includes one group of finely milled WC particles, ranging in diameter from about 1 μm to about 4 μm, with a mean ''effective'' diameter $D_{WC} \approx 2.5$ μm. The WC particles are generally polygonal or granular in shape. The second mode consists of significantly larger, quasispherical $A_6$ tool steel particles, ranging from about 20 μm (viz. *not* passing through a #600 mesh sieve) to about 38 μm (viz., passing through a #400 mesh sieve). Their mean diameter, $D_{A6}$, is approximately 27 μm.

This combination of particles provides the following benefits relative to simply using spherical particles of a single diameter:

1. For $\{D_{A6}/D_{WC}\} > 7$, the bimodal packing density is significantly higher.
2. The fine WC particles can fill the interstices between the larger $A_6$ particles.
3. Thus, the binder concentration is significantly smaller.
4. Consequently, there is less binder to be eliminated in the reduction furnace.
5. The mean shrinkage is smaller in the reduction, sintering, and infiltration steps.

6. Thus, the random noise shrinkage is smaller and accuracy is improved.
7. The smaller particles provide an improved surface finish.
8. The extremely hard WC particles improve insert abrasion resistance.
9. The $A_6$ tool steel provides good toughness, offsetting the brittle nature of WC.

Figure 2 schematically illustrates the benefits of a properly selected bimodal distribution on the final packing fraction, relative to using spheres of a single diameter.

It can be shown that the *maximum* possible packing fraction, $[F_{P,I}]^{max}$, for close-packed spheres all having the same diameter (i.e., a monomodal distribution) is given by the expression

$$[F_{P,I}]^{max} = \frac{\pi}{(3\sqrt{2})} \approx 0.74 \tag{1}$$

However, for a *bimodal* particle size distribution, the ratio, $R^*$, is defined by the expression

$$R^* \equiv \frac{D_{L,S}}{D_{S,L}} \tag{2}$$

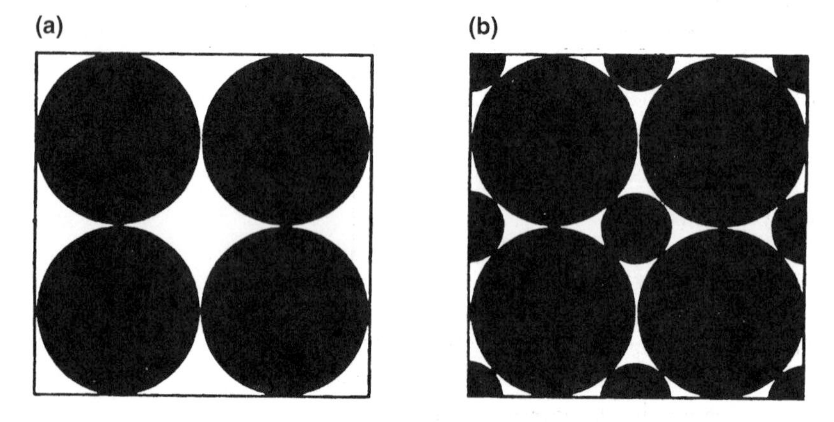


**Figure 2** The influence of (a) monomodal versus (b) bimodal particle distributions on final packing fraction. (From Ref. 1.)

where $D_{L,S}$ is the smallest diameter of the *large* particles and $D_{S,L}$ is the largest diameter of the *small* particles. Provided $R^* \geq 7$, then the smaller particles will be able to ''flow'' into the interstices between the large particles. This *small-particle flow condition* is key to obtaining higher values of the packing fraction. The effect of particle size ratio on the binary packing fraction is shown in Fig. 3.

It can also be shown that the maximum *bimodal* packing fraction, $[F_{P,\Pi}]^{max}$, is then given by

$$\begin{aligned}[F_{P,\Pi}]^{max} &= [F_{P,I}]^{max} + (1 - [F_{P,I}]^{max})[F_{P,I}]^{max} \\ &\approx 0.74 + (1 - 0.74)(0.74) \approx 0.93\end{aligned} \tag{3}$$

Here, the first term is the maximum packing fraction for the large spheres alone, and the second term is the maximum packing fraction for the small spheres in the remaining void space, provided they are small enough to flow into the interstices between the large spheres. To obtain high packing fractions, there are clearly advantages to using a bimodal distribution relative to a monomodal distribution. Extending this to trimodal or higher distributions would seem logical. However, resolution issues as well as economics definitely establish practical limits.

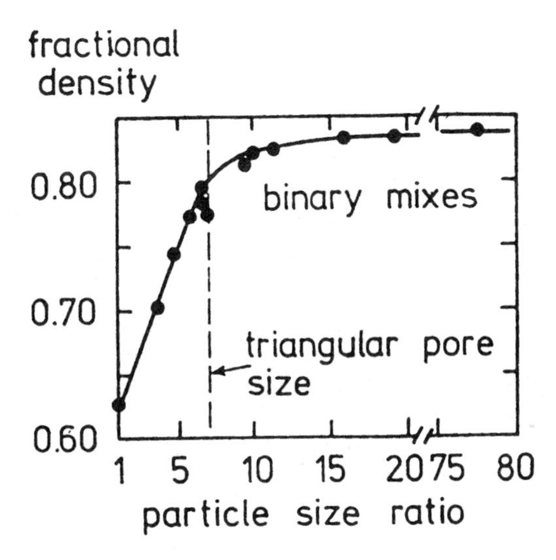


**Figure 3** The effect of particle size ratio on binary packing fraction. (From Ref. 1.)

To utilize a proper trimodal particle size distribution, the $R^*$ ratio must be applied *twice*. Furthermore, each individual particle distribution has a finite diameter span (viz. the ratio of the diameter of the largest particles in a single group to the diameter of the smallest particles in the *same* group). Unless very expensive and time-consuming separation techniques are utilized, it is difficult to obtain particles in a single group having less than a factor-of-2 diameter span.

Thus, the overall range of diameters required for a proper *trimodal* particle size distribution involves a cumulative diameter size factor $F \approx 2^3 \times 7^2 \approx 400$, where the $2^3$ factor involves the accumulated minimum diameter spans of all three modes, and the $7^2$ factor involves the $R^*$ ratio between the first and second modes, and then a second $R^*$ ratio between the second and third modes. This overall factor of about 400 in diameter leads to some important practical limitations. If the smallest particle diameter is about 1 μm, then the largest particles would have a diameter of about 400 μm. The problem with this approach is that 400 μm would then be the smallest positive feature that the system could replicate! As tooling inserts are often required to produce features as small as 100 μm, the 400-μm particles would seriously compromise overall tooling resolution.

Conversely, if we select the very largest particles to be 100 μm in order to achieve acceptable system resolution, then the smallest particles should be no larger than 0.25 μm. The twin problems which occur in this case are (a) cost and (b) agglomeration. In general, the smaller the desired particle diameter, the greater the powder cost, because extensive milling is often necessary. Also, complex and expensive methods of separating particles are required to differentiate between those falling within the desired size range and those falling outside that range. Furthermore, very tiny particles tend to agglomerate into larger multiparticle clumps, which then defeats the whole point of using trimodal distributions.

Finally, it is important to give the reader a sense of what has actually been *achieved* with real particles, relative to the *theoretical* maximum packing fractions for both monomodal as well as bimodal particle size distributions of perfect spheres. Measurements of the quasimono modal RapidTool process indicate average packing fractions of about 60%. This value is significantly below the theoretical maximum of 74%.

Some of the possible reasons are as follows:

1. The real particles are not perfect spheres.
2. The particles are *not* of uniform diameter.
3. The range of diameters from roughly 30 μm (not passing through

a #450 mesh) to about 75 μm (passing through a #200 mesh) causes imperfect packing and dislocations.

4. The 30-μm particles are too large to properly fill the natural interstices between the 75-μm particles.
5. Agglomeration of the smaller particles can also lead to imperfect packing.
6. Interparticle friction can result in void formation because the smaller particles are often unable to migrate into the interstices between the large particles.

In the case of the 3D Keltool process, which utilizes a bimodal particle size distribution, packing fraction measurements average ~70%. Although higher than 60%, this is also well short of the theoretical maximum of ~93%. It is likely that many, if not all, of the seven reasons noted account for this discrepancy. Lest the reader think that it must be utterly impossible to achieve packing fractions close to the theoretical maxima, it is worth noting that values as high as 71% have been achieved for monomodal distributions (1). Finally, McGeary produced a bimodal powder mixture with a packing fraction of 89% that still exhibited free-flowing behavior!

It is important for the reader to recognize that what is critical is not $F_P$ itself, but rather the quantity $1 - F_P$, which is directly proportional to the amount of binder to be eliminated. After the binder has been decomposed in the reduction process, a void space remains. The mean shrinkage is proportional to the extent of void space, because the sintered particles tend to compact toward full density by filling available interior volume. Finally, from Eq. (8) of chapter 4, the random shrinkage should be proportional to the mean shrinkage.

Because the random shrinkage limits the tolerances that such a process can satisfy per Eq. (9) of Chapter 4, insert accuracy is proportional to $1 - F_P$. Consequently, when $F_P = 0.60$, for the RapidTool process, the random shrinkage is proportional to $1 - F_P = 0.40$. For 3D Keltool, where $F_P = 0.70$, the random shrinkage is proportional to $1 - F_P = 0.30$. If values of $F_P$ as high as 0.85 could be achieved, the random shrinkage could be cut in half! This is the reason that high packing fractions are critical to improved core and cavity dimensional accuracy.

After the mixture of binder and $WC/A_6$ particles has cured in the RTV mold, the resulting ''green'' part is demolded. This operation must be performed carefully. Although the RTV mold is quite flexible, it can still generate considerable stress on thin sections. Because the green part is fragile, the yield of successfully demolded parts showing no damage from this step whatsoever is well below 100%. Techniques involving the use of air pressure on the RTV side and a vacuum on the green part side can help, but they are not universal solutions for all the various geometries encountered. Each new geometry can be an adventure unto itself.

After a properly demolded green part is placed in a reduction furnace, the ambient air is evacuated, and the furnace is then purged with nitrogen. For safety, the vacuum pumping/nitrogen purge sequence is performed *twice* to assure that the oxygen concentration is below 0.01% (i.e., 100 ppm). Next, a continuous flow of hydrogen is introduced into the furnace. The electrical heater elements are subsequently energized, and the electronically controlled furnace temperature, $T_F$, slowly ramps upward. When $T_F$ reaches about 350°C (~660°F), the hydrocarbon binder begins to decompose.

The dominant reaction is the reduction of carbon in a hydrogen atmosphere. The result is the production of methane ($CH_4$) according to the chemical reaction

$$C + 2H_2 \rightarrow CH_4 \tag{4}$$

At $T_F \approx 700°C \approx 1300°F$, the binder is almost fully decomposed except for very small amounts of carbonaceous residue. This residue acts like a form of ''glue,'' helping hold the green part together. Next, $T_F$ is raised to about 900°C (~1650°F). At this temperature, the $A_6$ steel particles undergo surface diffusion (3) at their respective points of contact. In the 3D Keltool process, the goal is to *avoid* liquid-phase sintering, as this leads to increased shrinkage.

The furnace is now allowed to slowly cool down over a period of about 20 h. The resulting article is essentially a porous skeleton consisting of about 70% interconnected WC and $A_6$ tool steel, with about 30% void space. After removing the skeleton from the furnace and completing a proprietary step intended to control infiltration at the active tooling surfaces, the article is again placed in a reduction furnace. This may be the same furnace used previously or another furnace specifically optimized for infiltration.

An excess of copper-alloy powder, relative to the void space within the skeleton, is also placed inside the furnace. A procedure similar to that described for the binder decomposition and sintering steps is initiated to ensure that hydrogen will not ignite in the presence of hot oxygen. This again involves twin vacuum pumping/$N_2$ purging sequences, followed by continuous $H_2$ flow. The furnace is then heated to $T_F \approx 1100°C \approx 2000°F$. At this temperature, the copper alloy becomes molten and infiltrates the porous skeleton under the influence of capillary forces. Properly infiltrated 3D Keltool parts are greater than 99% fully dense. The furnace's electronic temperature control system now initiates another cool-down over a period of about 24 h.

The cooled, copper-alloy-infiltrated, WC/$A_6$ tooling insert is then removed from the furnace. Next, the base is milled flat, eliminating excess copper-alloy infiltrant. The insert is then forwarded to inspection, where surface quality is assessed and critical dimensions are measured to assure that the part meets specification. If one or more dimensions do *not* meet specification, they may be machined if practical, or in some cases, it may be necessary to repeat the last two steps shown in Fig. 1. When all is well, the part proceeds to shipping. To save valuable time, it is returned to the customer by any of a number of ''next day'' airborne/courier services.

The major advantages of the 3D Keltool process are now listed. These advantages are relative to other rapid tooling processes that are currently commercially available. With the exception of items 3 and 6, they are *not* necessarily advantages relative to computer numerically controlled (CNC) and EDM core and cavity machining, which clearly set the current standards for production tooling.

1. The process has already been used to generate thousands of inserts over the past 20 years.
2. The inherent surface quality is quite good ($30 < R_A < 50$ μin.), due to the small WC particles, and can be polished to a mirrorlike finish (i.e., $R_A < 3$ μin.).
3. The thermal conductivity is better than conventional tool steel, due to the presence of about 30% copper. This can lead to shorter injection-molding cycle times and a corresponding increase in manufacturing productivity.
4. Abrasion resistance is very good, due to the great number of hard WC particles.
5. Tool life is excellent. Some Keltool inserts have achieved more than 3 million shots for unfilled thermoplastics such as polypropylene, ABS, nylon, and polycarbonate. Other inserts have achieved over 500,000 shots with *glass-fiber-filled* thermoplastics.
6. The process is indeed rapid. Regular turn around time is 4 weeks, from receipt of a valid .STL file until delivery of the core or cavity insert. Three-week delivery is available for an increased fee. *Two-week delivery*, or *''super-rush,'' can be expedited for a still higher fee.*

As a notable sidelight to illustrate the importance of rapid time-to-market, during the first quarter of 1997, roughly 80% of 3D Keltool customers chose the 2-week ''superrush'' schedule, in spite of the added cost. In many circumstances, such as those faced by the Project Widget team, saving a few weeks of product development *time* is clearly worth far more to many companies than the incremental fees for rapid delivery of tooling inserts.

The primary limitations of the 3D Keltool process are the following:

1. Obtaining truly accurate, dimensionally stable patterns. As good as RP&M processes have become, even SL ACES patterns, which have been the accuracy standard of the RP&M industry for some time, are often not capable of holding production tooling tolerances of ± 0.002 in. for dimensions greater than about 5 in.
2. Currently, the 3D Keltool process is limited to inserts that will fit in about a 4-in. cube. One source of this limitation is warping during the infiltration process. Long, thin, flat geometries are more problematic in this regard than short, thick geometries.
3. In addition to the pattern and warping issues noted, 3D Keltool accuracy is also limited by variations in shrinkage from one core or cavity to another, and even from section to section within a given core or cavity. This problem also increases with size.
4. 3D Systems bought the Keltool process in September 1996, and has recently licensed the process to other organizations. Repeatability, accuracy, and consistency have been problems. Although some core/cavity pairs fit beautifully at parting surfaces and shutoffs, some do not. Some inserts may require additional postmachining to offset these shortcomings. This, of course, begins to nibble away at both the time and cost benefits of the process. Unfortunately, some inserts are sufficiently warped that machining to flatness can cause other dimensions to fall outside their allowable tolerance. Finally, and not at all insignificant, if the error is on the lean rather than the proud side, correction by machining may not be possible, and the process may have to be repeated. This can significantly extend the delivery date to the point where much of the anticipated time savings are no longer realized.

## III. INVESTMENT-CAST RAPID PRODUCTION TOOLING

Production investment-cast tooling is based on the ability to quickly generate a computer-aided design (CAD) solid model of the mold geometry by taking the Boolean reverse of a digitally defined object. An example of this procedure is illustrated in Fig. 4 which shows the CAD design of the desired component

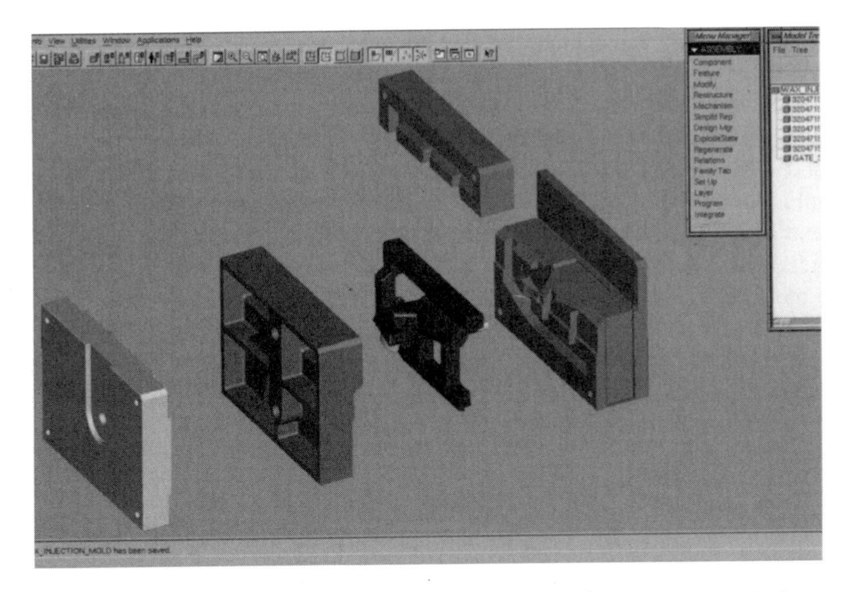

**Figure 4** CAD solid model of a part, and the core and cavity mold geometries resulting from a Boolean reverse of that CAD model.

upper right, and a corresponding Boolean reverse. The latter has been divided by a parting surface to form core (left) and cavity (right) mold sections. Early work on investment cast tooling was done by Denton (4) and is described in detail in Ref. 5.

The core and cavity patterns are produced using an RP&M system in a format suitable for the shell investment-casting process. An excellent example is the QuickCast™ build style developed by 3D Systems in 1992 and released commercially in 1993 (6). Metal castings are produced directly from the QuickCast patterns. After any necessary secondary machining, registration, and assembly operations are performed, the mold is ready for use.

Cast tooling is finding opportunities in a number of female cavity mold applications as inserts for the following:

1. Die casting
2. Rubber molds
3. Blow molding
4. Permanent molds
5. Plastic injection molding
6. Wax injection molds for investment casting

In the last application, the technologies have come full circle. Here, CAD is used to (a) generate a solid model of the part, (b) establish the parting surface, and (c) perform the Boolean reverse that establishes the core and cavity geometries. After shrinkage compensation, the resulting core and cavity solid CAD models are used to develop .STL files. An RP&M system then builds appropriate patterns of the core and cavity that will then be shell investment cast. The core and cavity patterns are investment cast, finish machined, aligned, and assembled in a tool base. The investment cast tool is then used to mold production quantities of wax patterns for investment casting.

To be competitive with conventional CNC and EDM mold-making practices, (a) the CAD design of the component and the mold, (b) the RP&M pattern fabrication, (c) the investment casting process, (d) the final machining steps, and (e) the tool assembly operations must *all* be

- Fast
- Accurate
- Economical

The design of the mold for investment-cast rapid production tooling employs practices similar to those encountered in conventional mold making, including the following:

- Defining the geometry of the production component
- Establishing a suitable parting surface or surfaces
- Defining the required core and cavity geometries
- Defining any side cores, loose inserts, injection systems, ejection systems, and so forth

However, a fundamental difference is that for investment-cast tooling, *all* the geometries are generated using solid CAD. In most cases, the customer supplies a CAD solid model of the desired component. This digital information then becomes the basis of all subsequent operations. The digital creation of the mold components is accomplished by taking one or more Boolean reverses of an appropriately shrinkage-factored CAD solid model of the desired final component.

The casting engineer must initially evaluate the production component geometry, as well as the intended wax and metal gating systems. All of the various forces that affect the volumetric shrinkage of the pattern wax and the metal casting are considered before determining the relevant mold shrinkage factors. The number of factors used will range from 1 for small castings to more than 100 for large castings with numerous physical attributes.

Applicable shrinkage factors are used to develop a ''shrinkage-compensated solid CAD model'' of the final component. It is this shrinkage-compensated digital model that is ultimately used to establish the Boolean reverse, the parting surface(s), the core and cavity geometries, the .STL files, and, finally, the RP&M core and cavity patterns that will ultimately be investment cast. All features required in the final cast configuration must be incorporated into the shrinkage-compensated solid model of the core or cavity. This includes fillets, rounds, identification numbers, part markings, customer and foundry trademarks, and so forth. Any casting enhancements such as (a) grinding stock, (b) vents, (c) drains, (d) wax gates and runners, and (e) liquid metal gates and runners are CAD modeled at this time.

The heart of the investment-cast rapid production tooling (INC–RPT) process is the development of an appropriate *shrinkage-compensated solid CAD model* of the final component. If this is done correctly, generating solid CAD models of the mold core and cavity are relatively straightforward. A block shape is defined in CAD, having extents in $X$, $Y$, and $Z$. These must be sufficient to ensure that the shrinkage-compensated solid CAD model of the component can fit inside the block, with enough room to spare in all directions. This is important because the final core and cavity must be strong enough to provide the required life of the production tool. Once this has been done, the shrinkage-compensated solid CAD model of the component is subtracted from the block. The result of this reverse should be the desired geometry of the final mold.

Various CAD vendors have developed software packages for the specific purpose of efficiently generating mold geometries. Parametric Technologies Corporation has a product for this purpose called ProMoldesign™. The examples shown in Fig. 5, as well as others in this section, were generated using this software module. The designer is assisted through real-time feedback relating to potential part/mold locking conditions. The opportunity is then available to change the configuration of the final component in an effort to simplify the design of the mold or to create an additional core or insert in order to accommodate part extraction.

Once the mold components have been defined, a determination is made regarding those surfaces that require excess machine stock. Candidate locations are those where as-cast surfaces will not satisfy the functional mold requirements for surface finish, flatness, or dimensional accuracy. This may be due to the variability of the pattern-making process, random-noise shrinkage in the investment-casting process, or a combination of both. Typical examples that require final machining would be parting surfaces and shutoffs.

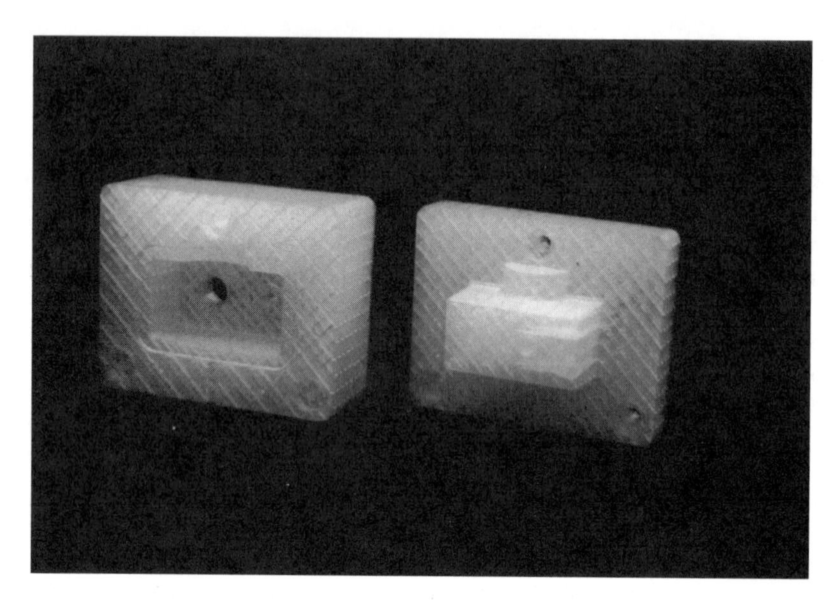

**Figure 5** An SL QuickCast pattern to be used for investment casting.

Not all mold components will be cast. Where a feature can be formed by a process which itself is faster, more accurate, or less expensive, that method should be used. An example would be a round core pin machined on a lathe from standard bar stock. Here, the pin can be machined much faster and at significantly lower cost than it could be investment cast. Nonetheless, even though the component is made by another process, it is still included in the CAD solid model. This allows the mold to be ''operated'' and ''cycled'' in a virtual manner.

Through computer simulation, it is now possible to ''assemble'' the mold components, ''inject wax'' into the mold, and ''view'' the filling action. One can also ''disassemble'' the mold in the correct operational sequence, ''extract'' the solidified wax pattern, and then evaluate its features for completeness. Again, all of these tasks can now be accomplished digitally.

It is well known that the best time to catch an error is at the earliest possible point of discovery. The virtual world provides an excellent opportunity to efficiently uncover such problems *before* expensive and time-consuming hardware changes must be made.

Once defined, the individual CAD models of the mold patterns must have their cast shrinkage evaluations performed and the relevant shrinkage-compensation factors applied. For the case study which follows, the time required to accomplish (a) the CAD solid modeling, (b) the complete shrinkage-compensation factoring, and (c) all the associated process engineering tasks up to, but not including, the generation of the RP&M patterns was just *5 calendar days*.

Throughout this discussion, all time intervals will be given in elapsed or ''calendar'' time. This is quite different from what might be referred to as ''stopwatch'' time, where only the time required to accomplish specific actions or tasks is counted, with any queue time *between* tasks being ''conveniently omitted.'' Ultimately, it is the *total elapsed time* that really matters when faced with a deadline; hence, it is calendar time that will be reported. Of course, the calendar time will always be longer than the stopwatch time often reported by others. Therefore, the results may not appear as dramatic. However, the data are indicative of what a user can realistically expect. Furthermore, just as it is important to reduce the time for each individual step, analysis of total elapsed time will probably point out other intervals where time can also be saved.

The core and cavity patterns for this case study were created using the QuickCast build style, which allows cured photopolymer to function as an expendable pattern for the ceramic shell investment-casting process. QuickCast establishes the pattern geometry with a thin skin (~1 mm thick), supported by an interconnected quasihollow hatch structure. This build style allows the SL pattern to successfully emulate the petroleum-based wax patterns used in conventional investment casting (7). Figure 5 shows one of the QuickCast patterns used in this study.

The QuickCast process initially requires checking that the pattern is well drained and free of internal, uncured liquid resin. After postcure, the next step involves filling the vent and drain holes that were intentionally generated to evacuate uncured liquid resin from within the pattern. Filling the holes can be done either with investment-casting wax or photopolymer resin thickened to a pastelike consistency using fine powder ground from previously solidified photopolymer. The pattern should then be tested to ensure that no openings exist. This is best done with about 4–5 psi (~0.3 bar) of positive pressure, followed by drawing a mild vacuum (~0.7 bar absolute pressure). In either case, any leakage indicates the presence of one or more holes.

Once the pattern has been properly sealed, it is then connected to its associated gating system. The pattern and gating are subsequently encapsulated in a multilayer ceramic ''slip'' and refractory grain coating. This coating is allowed to air-dry, after which the entire assembly is placed in a furnace preheated to about 1000°C (~1800°F). The original QuickCast 1.0 software version released in 1993 exhibited a triangular internal structure roughly 65–70% void. QuickCast 1.1, released in 1995, had a square internal structure about 80–85% void. Many outstanding investment castings (8) were generated from QuickCast 1.1 during the period from 1995 to early 1997. Recently, 3D Systems released QuickCast 2.0, with a hexagonal internal structure having void ratios in the 88–92% range. Based on the work of Hague and Dickens (9), QuickCast 2.0 patterns produce less than one-third the shell stress of QuickCast 1.1 during pattern burnout, significantly reducing the probability of shell cracking.

Flash-firing the mold eliminates the photopolymer pattern/gating system by burning the hydrocarbon-based resin in the presence of not less than 10% free oxygen, at temperatures around 1000°C. The result is a hollow, disposable, ceramic receptacle for molten metal. The QuickCast pattern will indeed expand when heated, but the expansion forces take the path of least resistance, collapsing the internal structure, which weakens significantly above the resin glass transition temperature, at about 70°C. The result is a significant reduction in ceramic shell stress. The patterns for this project were built on an SLA 350/10, using CibaTool SL 5190 epoxy resin. All QuickCast patterns were built, drained, postcured, vents/drains filled, patterns pressure checked, and exterior surfaces carefully sanded and finished, within an additional 3 *calendar days*.

A quality assurance/dimensional verification of the patterns was then performed. This was accomplished with the help of the SolidView™ software package, enabling extraction of the dimensional extents of major part features from the .STL file used to build the patterns. After all preparations were completed, the patterns were investment cast using standard QuickCast procedures. Including the time required for gating removal, heat treating, straightening, and various finishing operations, the investment casting process took another 5 *calendar days*.

The investment-cast mold components are shown in Fig. 6. The castings were inspected to determine the final dimensions based on the *actual* shrinkage values. Incorporation of registration pins, hold-down bolts, finish machining of parting and shutoff surfaces, retaining plates, cavity polishing, and mold venting took 2 *additional calendar days*.

Wax patterns were produced from the investment-cast rapid production tool. These patterns were dimensionally verified before committing to production investment casting. Fully functional, production investment castings, including heat treating, straightening, nondestructive testing, and dimensional verification of the final castings, were achieved in 5 *calendar days*.

**Figure 6** Investment-cast mold components.

The time sequence for the fabrication of INC–RPT, including investment casting the production components, was as follows:

1. CAD solid modeling of component, mold, and engineering tasks — 5 days
2. QuickCast mold pattern generation — 3 days
3. Investment casting of the mold components — 5 days
4. Incorporating related features into final mold, assembly, and test — 2 days
5. Molding the required wax patterns, and investment casting — 5 days

Total calendar time: CAD to production aluminum castings — 20 days

Figure 7 shows the assembled INC–RPT. Figure 8 shows one of the production wax patterns molded with the tooling illustrated in Fig. 7.

Rapid production tooling efforts based on applying RP&M technology to the venerable shell investment casting, or ''lost wax'' process, must certainly be considered a work-in-process. The technique potentially represents yet another means of satisfying the worldwide demand for reduced time-to-market. The current process limitations are primarily (a) CAD, (b) RP&M

**Figure 7** Assembled INC–RPT.

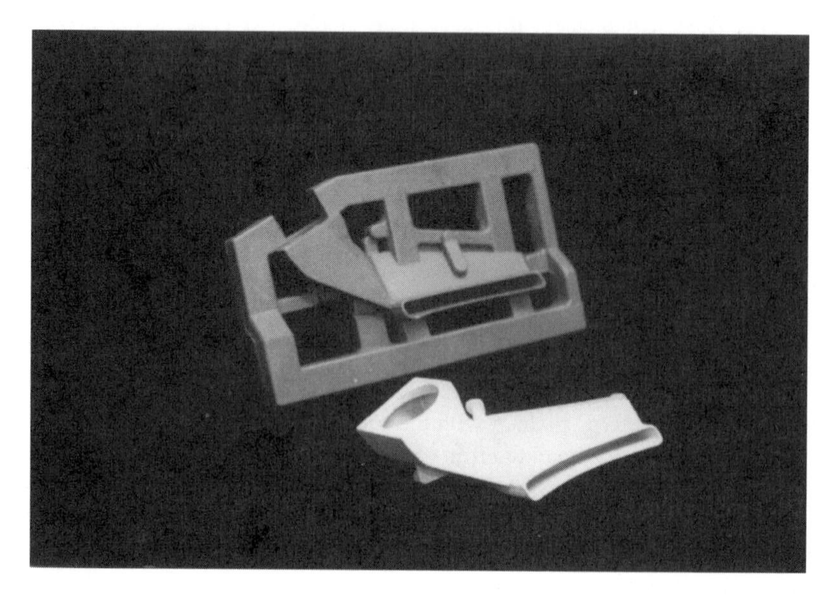

**Figure 8** Wax patterns molded from IC–RPT.

pattern accuracy, and (c) nonuniform shrinkage in the shell investment-casting process. Without careful attention to detail, these limitations could restrict the scope of INC–RPT. With the above limitations noted, when the production component requirements, coupled with critical time-to-market pressure fall within the capabilities of INC–RPT, a meaningful market segment will likely develop. In fact, more than a dozen INC–RPT tools have already been built, assembled, tested, and put into *production* within the past year by Solidiform, Inc. (Fort Worth, TX).

When evaluated against satisfying urgent requirements with respect to *time*, the procedure is clearly worth pursuing, as indicated by the case study described herein. *Going from a CAD solid model to fully functional production aluminum castings in less than 3 weeks* is certainly extraordinary. With proper implementation of the process by qualified personnel, working within the scope of the constraints noted, the acceptance and advancement of INC–RPT is likely to grow.

## IV. RAPID PRODUCTION TOOLING FOR PRECISION SAND CASTING

The advent of chemically bonded sand has brought a new term and a new capability to the world of the foundry. The term is *precision sand casting*. As the name implies, the process yields castings with finer surface finish, more intricate detail, and significantly higher-dimensional accuracy than previously possible with conventional green sand casting. Chemically bonded sand can replicate a surface quickly, accurately, and economically. This enables RP&M-generated tooling solutions that can satisfy especially time-sensitive cast-metal requirements.

The compatibility of chemically bonded sand with SL ACES patterns is enabling the production of as many as *1000 castings from a given configuration*. This unique tooling approach has already been successfully applied for both short-run prototype and long-run production requirements.

In its simplest form, the chemically bonded sand approach involves a sand mixer that coats very fine sand particles with a catalyst. This operation is accomplished in isolation from a second mixing operation that coats similar sand particles with a binding agent. Then, the catalyst-coated sand particles are brought together with the binder-coated sand particles in a high-speed mixing cone. Here, the two types of coated sand particles come in contact in a continuous stream. The output of the mixing cone is directed over a pattern set. The combined sand mixture is then tucked and hand compacted against the pattern, which is held by a rail set in the $X$–$Y$ plane. Figure 9 shows the mixing cone, with the resulting stream of mixed binder- and catalyst-coated fine sand particles being directed onto an ACES pattern.

The catalyst-to-binder ratio establishes the available ''working time'' of the sand before it takes an initial set. When first mixed, the sand is very fluid and is easily directed into the cope (top) and drag (bottom) pattern boxes. After a given amount of time, based on the sand volume and the catalyst/binder ratio, the mold will exhibit sufficient strength to allow inversion and pattern withdrawal without damaging the cured sand mold, provided reasonable care is exercised.

Once fully prepped, the two mold sections are closed against one another and clamped with sufficient force to withstand the hydrostatic pressure of the

**Figure 9** Precision sand-casting mold formation.

molten metal during the pouring operation. The binder holding the sand particles together, at the interface between the cast metal and the precision sand mold surface, is subsequently broken down by the high temperature of the molten metal. This results in a loose sand envelope adjacent to the casting. This ''thermal debinding'' facilitates the removal of the casting and its associated gating system from the mold with minimal stresses being imposed on thin or delicate cast sections.

Fortunately, the precision sand-casting process can produce quality sand castings quickly. For many simple configurations, it has already been demonstrated that it is possible to close the mold, pour molten metal, cool down, and remove the solidified metal casting within 1 h. Obviously, larger and more geometrically complex parts may take somewhat longer.

Although the thermal debinding mechanism does greatly assist casting removal, it does, of course, destroy the mold. In that sense, precision sand casting is similar to investment casting: both processes provide *one casting per mold.* However, it may take 5–10 *days* to create an investment casting shell from a single pattern. Furthermore, the pattern itself is eliminated in the investment-casting process. With precision sand casting, the same pattern may be used over and over again, and the mold can be produced in a matter of *hours*.

Virtual pattern making is not a term of the future, it is a fact now. The ability to utilize the skill sets of a journeyman pattern-maker to guide the construction of precision sand tooling through the computer is becoming less rare. Designing a pattern in a CAD environment employs procedures similar to those used in conventional pattern making; specifically, determining the parting surfaces, establishing the core prints, defining core boxes, and so forth.

The tooling for the precision sand case study shown in Fig. 10. involves a single impression cope and drag plate with its associated core boxes. All tooling components were modeled in solid CAD, including major portions of the gating systems.

From the solid CAD model, the primary parting surface is defined and the CAD model is split. Part features that will be formed by secondary cores are identified. Appropriate core prints are also CAD modeled for the respective cores. The core print, as well as with the core itself are extracted from the model as a single entity. This is illustrated in Fig. 11.

At this point, a core box can be modeled around the core and core print. This process is repeated until all cored areas are described. The sand mold, with all its cores in place, can be simulated in the computer. Finally,''molten metal'' can be ''poured'' in the computer simulation, and the resulting ''casting'' can be ''extracted'' and ''inspected.''

**Figure 10** Rapid tooling for precision sand casting, including the cope and drag plate and associated core boxes.

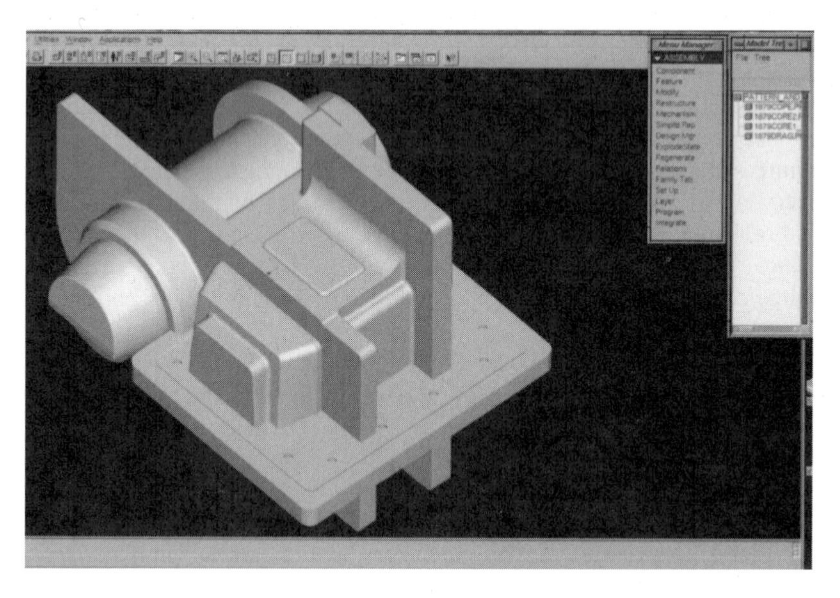

**Figure 11** Core and core print extracted from CAD model.

Once the design has been achieved, the individual components are generated on an SLA using the ACES build style. All tooling components are currently built using a 0.004-in. (~100 μm) layer thickness for maximum surface resolution and accuracy. Nonetheless, the components still require some benching prior to mounting and assembly, in order to eliminate "stair-stepping" on inclined surfaces. When thinner layers or advanced techniques such as "meniscus smoothing" become available, the improved surface quality of inclined or compound curved surfaces will greatly reduce the amount of benchwork. In turn, this will further accelerate the entire process.

Precision sand casting requires no external packing, pounding, or tamping. Consequently, the fine sand/binder/catalyst mixture can be molded against an ACES part with very little abrasion. As a result, there is almost no degradation of the ACES patterns during the sand-filling, mold-curing, or pattern-extraction steps. A seal coat of paint applied in a light color is suggested to further aid in the visual inspection of abrasion on the active tooling surfaces.

The ACES patterns have proven to be extraordinarily robust when used in a production mode. Some configurations have yielded over 1000 precision sand molds *without* any signs of wear. Obviously, care must be used in molding, pattern extraction, and general handling to allow for the reduced strength and impact resistance of cured epoxy resins relative to either aluminum or steel tooling. Experience to date indicates that tools fabricated in this manner certainly require care in their use, but, of course, this is true for any precision tooling.

For the case study described herein, the sequence of events and the time required to develop "Precision Sand-Cast Rapid Tooling" is listed. Note that this total elapsed *calendar* time includes not only the first article production casting, but weekend time as well.

| | |
|---|---|
| 1. Generating a solid CAD model of the casting from 2D customer data | 5 days |
| 2. Solid CAD modeling of the tooling, as well as associated engineering | 10 days |
| 3. Building the ACES patterns/core boxes | 10 days |
| 4. Bench finishing and assembling the tooling components | 5 days |
| 5. Producing the "first article casting" and performing QA inspection | 5 days |
| Total *calendar time* from customer 2D data arrival until delivery of the first article casting | 35 days |

The development of the practices and procedures needed to extract the greatest amount of *time* from the process while still delivering quality castings at a favorable *cost* continues. By combining the technologies of CAD, RP&M, and precision sand casting, it is now possible for customers to receive aerospace quality castings in quantities from 1 to 1000 in a time frame that would have been considered utterly impossible just 5 years ago.

Without question, the manipulation of digital data to produce tooling is the wave of the future. The prospect of being able to generate tooling with a computer-controlled additive system is truly fantastic. The word ''precision'' in the term precision sand casting takes on additional significance when augmented with the capabilities of solid CAD modeling and an accurate RP&M technique such as SL. The applications that can be addressed with these technologies appear limited only by our collective imaginations.

## REFERENCES


1. R German. Particle Packing Characteristics. Princeton, NJ: Metal Powder Industries Federation, 1989.
2. R McGeary. Mechanical packing of spherical particles. J Am Ceram Soc 44: 513–522, 1961.
3. R German. Powder Metallurgy Science. 2nd ed. Princeton, NJ: Metal Powder Industries Federation, 1994, pp. 242–267.
4. K Denton, P Jacobs. QuickCast and Rapid Tooling: A case history at Ford Motor Company. Proceedings of the SME Rapid Prototyping and Manufacturing '94 Conference, Dearborn, MI, 1994.
5. K Denton. Hard tooling applications of RP&M. In P Jacobs, ed. Stereolithography and Other RP&M Technologies. Dearborn, MI: SME Press/New York: ASME Press, 1996, pp. 293–315.
6. P Jacobs. The Development of QuickCast In: P Jacobs, ed. Stereolithography and Other RP&M Technologies. Dearborn, MI: SME Press/New York: ASME Press, 1996, pp. 183–207.
7. L Andre, L Daniels, S Kennerkecht, B Sarkis. QuickCast™ foundry experience In P Jacobs, ed. Stereolithography and Other RP&M Technologies. Dearborn, MI: SPE Press/New York: ASME Press, 1996, pp. 209–237.
8. P Blake, O Baumgardner. QuickCast applications. In: P Jacobs, ed. Stereolithography and Other RP&M Technologies. Dearborn, MI: SPE Press/New York: ASME Press, 1996, pp. 239–252.
9. R Hague, P Dickens. Stresses created in ceramic shells using QuickCast models. Proceedings of the 5th European Conference on Rapid Prototyping and Manufacturing, Helsinki, 1996, pp. 15–30.

# 6
# Nickel Ceramic Composite Tooling from RP&M Models

**Sean Wise**
*CEMCOM Corporation*
*Baltimore, Maryland*

## I. INTRODUCTION

A matched die mold fabrication technique is discussed where nickel is electroformed over special tool mandrels made by rapid prototyping and manufacturing (RP&M) methods. The resultant nickel shells are then captured in a standard pocketed mold frame using a high-strength chemically bonded ceramic (CBC) to secure the shell to the frame. The resulting nickel ceramic composite (NCC) mold has a high-tensile-strength, abrasion-resistant surface, coupled to the high-compressive-strength ceramic backing which provides support and mechanical load transfer to the mold frame. The match of the ceramic's thermal expansion coefficient to that of nickel, along with the net-shape forming characteristics of both materials help maintain an effective bond and precise location of the tooling components. This method was developed to produce precise, high-quality fully functional tooling capable of intermediate volume production runs in less than half the lead time of conventional machined metal tooling. This chapter describes the tool-fabrication method through case studies undertaken as part of the development effort, as well as the molding performance of the tools.

## II. RAPID TOOLING

The model building industry has been revolutionized with the growth and implementation of three-dimensional computer-aided design (3D CAD) tools coupled with RP&M model building methods. In recent years, development emphasis has shifted toward methods to create part-specific *manufacturing* hardware such as tooling just as rapidly as plastic models. If this can be done, then RP&M can become an integral part of the entire manufacturing process. Additive processes are attempting to directly or indirectly produce such tooling. These include powder metal methods (1), cast metal (2), and metal deposition (3). In order to have a major impact on the very long lead items in an original equipment manufacturer's (OEM) product-development cycle, rapid tooling methods must address the tooling needs for large parts, as this is where the potential benefits are greatest, and large tools are usually the pacing items in product-development programs, as shown in Fig. 1.

If one considers that benching, fitting, and finishing represent more than a third of the fabrication time in conventional machined metal tooling, *a truly rapid process must produce an accurate tool that requires a minimal amount*

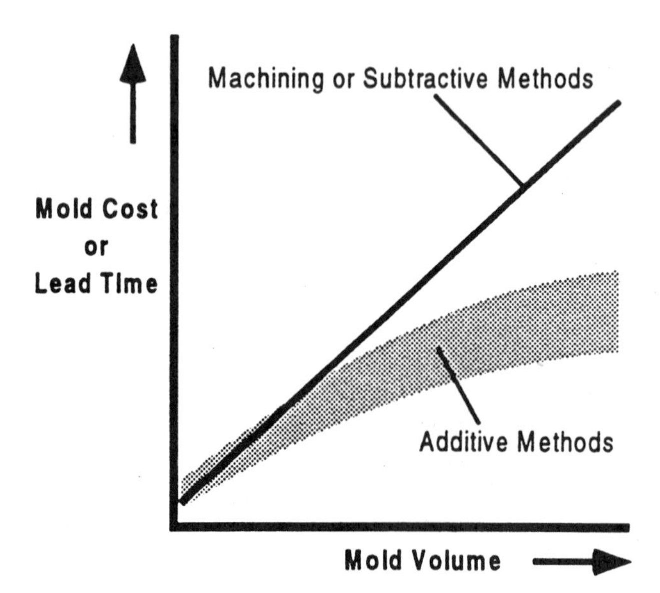


**Figure 1** Comparison of mold cost/lead time versus mold volume for tools made via additive or subtractive methods.

*of finishing or bench work.* A truly rapid tooling process must therefore have the following features:

1. Yield precise geometry so that rework associated with fitting mold components does not consume a significant portion of the time saved using an RP&M method
2. Provide a hard, durable surface with a good finish, directly
3. Not be limited in the *size* of molds that can be produced
4. Capable of high production runs and rates
5. Not be limited with respect to part features or geometry
6. Able to be produced in less than half the lead time of conventional machined steel tooling

One additive build method that can transfer geometry precisely from plastic RP&M models is nickel electroforming. If this process is combined with a high-strength backing material and standard mold components, tooling can be produced that meets all or nearly all the criteria defined. The tooling system presented in this chapter is called "nickel ceramic composite" tooling or simply NCC tooling.

## III. NCC TOOLING

Nickel ceramic composite tooling utilizes an electroformed nickel shell as the hard active surface of a mold, a high-strength ceramic as a backing for support, and a standard steel mold frame for containment. This is shown in Fig. 2. What makes this system unique is the way in which these three materials

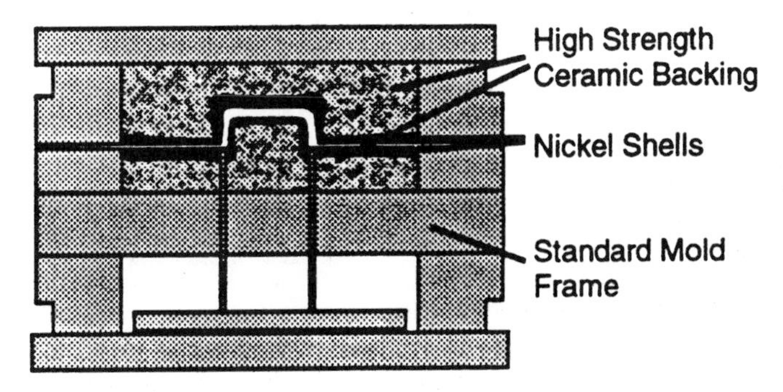


**Figure 2** General layout of a NCC tool showing the use an electroformed face that is coupled to a mold frame with rigid ceramic backing.

function together. The nickel face is a high-tensile-strength, abrasion-resistant surface that provides mechanical integrity for the most highly stressed areas of the mold. When formed, it will accurately reproduce even the finest surface detail or finish. The high-compressive-strength backing is an easily cast material and, as such, it can fill in behind the nickel shell to provide uniform support while saving both time and money. The pocketed metal frame provides containment and alignment of the nickel and ceramic elements. All three of these materials are quite stiff and have nearly identical coefficients of thermal expansion (CTE), thus the forming process for the nickel and the ceramic takes place with very small dimensional change. Furthermore, during the natural temperature cycles encountered in injection molding, all three elements of the system "move together," greatly reducing any tendency toward delamination.

For the system to function efficiently under injection-molding conditions, the backing must effectively transfer mechanical and thermal loads away from the nickel surface. This requires a very rigid backing. Figure 3 shows how the stress in the nickel and the displacement of the mold surface varies with the backing material modulus. These results are from a finite-element analysis of a geometry very similar to the cavity shown in Fig. 2, loaded with a pressure of 400 bar (5900 psi). The nickel shell must bear a greater portion of the load if the backing has a low modulus. On the other hand, if the backing is more rigid, the stress will be transferred more effectively to the mold frame. A higher backing material modulus results in less stress on the nickel shell and significantly lower displacements of the mold surfaces.

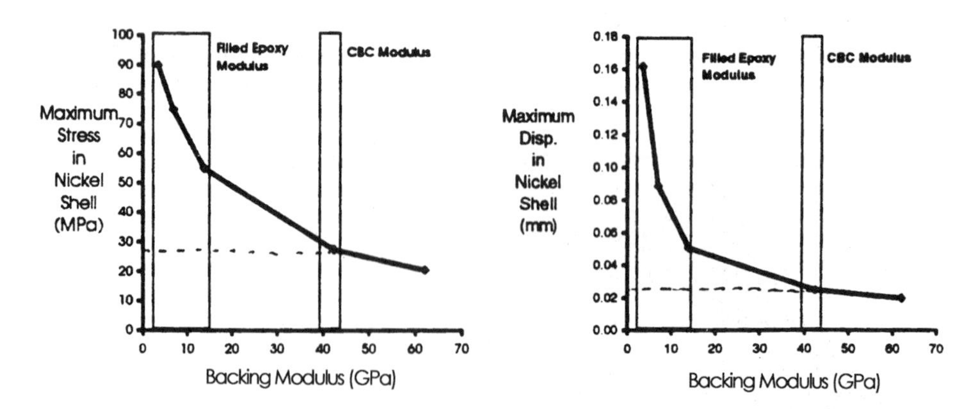


**Figure 3** Maximum stress and maximum displacement in nickel shell as a function of backing material stiffness. Note how stiffer backing reduces both nickel shell stress and displacement.

The difficulty with a very stiff backing is that any dimensional changes between the backing and the shell, such as those caused by differences in CTE, or shrinkage of the backing after it has solidified, will result in a substantial shear stress at the interface. The COMTEK 66 chemically bonded ceramic, with its CTE closely matched to Ni (13.9 ppm/°C versus 13.5 ppm/°C) also provides very low shrinkage during cure (±0.02%), resulting in an excellent support material. (See Table 1.) Its room-temperature forming characteristics are also important, as are its thermal conductivity. Although the ceramic's heat-transport properties are not as good as most metals, it is more than twice as good as aluminum-filled epoxies [2.5 W/m K versus 1 W/m K (4)] and ten times better than unfilled epoxies [0.2 W/m K (5)]. In addition, the properties of the ceramic backing do not change when exposed to temperatures up to 400°C (~750°F) (6).

## IV. NCC TOOLS BASED ON STEREOLITHOGRAPHY MODELS

Recognizing the potential of the nickel ceramic combination, CEMCOM examined the use of RP&M mandrels as the geometric basis for this hard tooling method. The process was used to form a test injection mold of an ice scraper (Fig. 4) where the electroforming mandrel was an stereolithography (SL) model. This work was performed in conjunction with Pennsylyania State University, Erie (3). The model was plated using a high-speed nickel process that built up the required metal thickness in less than 4 days.

However, the electroforming conditions resulted in significant deformation of the parting plane surface, and the nickel shutoff areas had to be ma-

**Figure 4** Sketch of ice scraper part made in conjunction with Penn State University, Erie.

**Table 1** Properties of Electroformed Nickel and COMTEK 66 CBC Tooling Compound

| | Electroformed nickel | | COMTEK 66 | |
|---|---|---|---|---|
| | English units | S.I. units | English units | S.I. units |
| Tensile strength | >70 kpsi | >500 MPa | | |
| Compressive strength | | | 50 kpsi | 350 MPa |
| Flexural strength | | | 6 kpsi | 42 MPa |
| Elastic modulus | 29 mpsi | 200 GPa | 5.4 mpsi | 37 GPa |
| Hardness | $R_c$ 20 | — | $R_b$ 65 | — |
| Coefficient of thermal expansion | 7.5 × $10^{-6}$/°F | 13.5 × $10^{-6}$/°C | 7.7 × $10^{-6}$/°F | 13.9 × $10^{-6}$/°C |
| Thermal conductivity | 468 BTU in./ft$^2$ h °F | 67 W/m K | 17 BTU in./ft$^2$ h °F | 2.5 W/m K |
| Specific heat | 0.11 BTU/lb °F | 450 J/kg °C | 0.19 BTU/lb °F | 787 J/kg °C |
| Shrinkage | Nil | Nil | ±0.2 mil/in. | ±0.02% |
| Max. operating temperature | 500°F | 260°C (+)[a] | 400°F (+)[a] | 200°C (+)[a] |

[a] Above 200°C (400°F) shrinkage increases. COMTEK 66 has been used in plastic part fabrication at up to 315°C (600°F) and metal fabrication at 540°C (1000°F).

chined in order for the tool to close properly. Even so, once the parting surfaces were fitted, the tool was set up in a press and more than 1000 parts were molded without signs of wear. This test showed the potential of the process, but the stability and accuracy of the SL mandrel had to be improved if the NCC process would ever satisfy the requirements for rapid tooling.

One way of avoiding the stability problems of the SL mandrel is to only use it as the basis for a secondary pattern that is compatible with the electroforming process. CEMCOM fabricated a tool in conjunction with the Queensland Manufacturing Institute (QMI) and Marky Industries following this process. This tool and the parts made in it are shown in Fig. 5. The electroforming mandrels were made from a standard tooling epoxy, then plated, and the nickel shells subsequently separated. The shells were brought to CEMCOM, backed with COMTEK 66 ceramic, and returned to QMI. The outer surfaces of the ceramic were then machined flat, fitted to a base which contained alignment features, and the two halves were fit together and finish machined to make a close tolerance tool shutoff. The mold was then run to show that it was capable of generating an injection-molded part. Although high-quality molded pieces were again produced, the amount of fitting and alignment was very time-consuming. Excessive fitting and alignment needs to be eliminated if the process is to be fast and competitive.

**Figure 5** Nickel ceramic composite mold fitted to a mold plate along with molded parts.

## V. INTEGRATION OF TOOL FORMING WITH RP&M MANDRELS

Having established that the combination of nickel and chemically bonded ceramic works well, the focus of this project concentrated on integration of the tool-forming steps with RP&M mandrel making (7). It is one thing to have a process which is net shape relative to a model, but it is quite another to have one where the model's geometry will not be compromised by the processing to which it is exposed. In essence, the process requires that a piece of plastic, which is not as stable or as accurate as a machined piece of metal under the best of circumstances, provide the precise geometry needed in a high-pressure forming mold. Earlier experience with RP&M mandrels used to make the PSU ice scraper showed these limitations. The electroforming process takes place under water, well above room temperature. As the nickel deposits on one side of the model, it seals the surface to moisture penetration. If just one side of the RP&M mandrel is being used to form the tool surface, then bending in the pattern will take place as one side swells while the other side does not. Once this happens on core and cavity patterns, the parting plane is compromised, so the tool is both difficult to seal and the part thickness is not precisely controlled.

If the part model is designed so that the core and cavity geometry are each attached to a common parting plane, then the tendency to distort will be minimized because both will be sealed by the nickel. If any swelling takes place, it will be very nearly identical on both sides. In addition, even if minor distortion does take place, the connection between core and cavity ensures that one side will follow the other. Besides assuring better parting line accuracy, the two-sided-model concept provides substantial benefits with respect to core and cavity alignment during tool assembly.

The process for using a two-sided RP&M mandrel as the basis for an NCC tool is illustrated in Fig. 6. It begins with the electronic 3D solid CAD model of the part. A part designer in conjunction with a tool designer analyzes the CAD model, and the parting surface is defined around the perimeter of the part. This surface, and the part model itself, is expanded in a linear fashion in the mold opening or $z$ direction. The amount of $z$ expansion should be chosen to provide a model with good stability for the plating conditions. If the part is edge gated, the runner and gate should be laid out on the parting plane to ensure that there is sufficient nickel over these surfaces. The parting surface should also have provisions for alignment to the mold frame, as this will facilitate assembly and backing in the later stages of toolmaking.

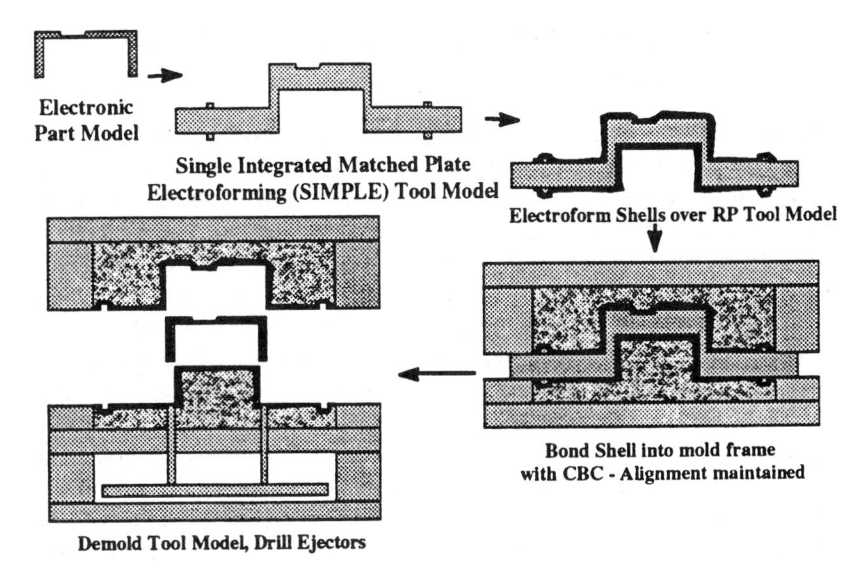


**Figure 6** Schematic of tool-fabrication process using a two-sided single model with an integral parting plane.

The particular model used in this test had a buss bar/clamp ring attached to its perimeter. This ring had three purposes: (a) create electrical contact around the perimeter of the model, (b) clamp the model to prevent bowing or distortion around the edges, and (c) provide a means of attachment to the mold frame. Because the clamp ring attaches to a machined recess in the mold frame, these two rings need to be precisely aligned relative to one another, or a shift will occur in the core relative to the cavity during final assembly.

The nickel shells are electroformed over the mandrel using plating conditions that minimize dimensional changes of the RP&M material. The amount of nickel needed on the model will be dependent on factors related to the stresses incurred during part forming, the number of parts needed, and so forth. With this method, the nickel thickness required is somewhat less than ordinarily used for electroformed shells because the ceramic backing will provide the needed support in the thinner areas and the shell does not need to be removed from the model until *after* it has been backed. This approach allows shells to be made in days rather than weeks.

The RP&M mandrel, with the electroformed nickel on both front and back surfaces, is attached to the mold frame using the clamp-ring buss bars.

The ceramic is then vacuum cast through a small opening in the back of the frame. The material must harden overnight before the mold is flipped and the opposing side can be cast. Once the opposing side hardens, the two halves are separated, removing the RP mandrel from the core and cavity. This step is then followed by a cursory inspection. Next, the ceramic is hydrothermally cured and measured and the fit of the core and cavity checked. The nickel surfaces can be cleaned or polished at this stage if needed. RP models normally have roughness and/or stair-stepping on vertical walls. Much of this should be sanded smooth prior to electroforming, but there may be inaccessible areas on the mandrel which are difficult to finish. Given the ability of the nickel to pick up very fine detail, these imperfections transfer to the nickel surface. The work required to clean this up is usually minimal, as the most recessed areas on the models now stand proud on the tool surfaces. Sampling of the mold can be performed without any finishing of the nickel tool face, but the rough surfaces with a low draft angle may not release well. In order to get the tool ready for molding parts, holes for ejector pins need to be drilled through the nickel, ceramic, and mold frame. CBC is sufficiently machinable that holes for ejector pins larger than 1/8 in. in diameter can be drilled accurately. Holes smaller than 1/8 in. in diameter may require a bushing. The runner is also extended through the buss bar ring and into the mold frame to connect it to the sprue, enabling the tool to mold parts.

## VI. A SMALL DEMONSTRATION TOOL

The model build approach outlined in Figure 6 was tested on a part geometry and SL mandrel supplied by Doug Van Putte formerly of Kodak (8). This part, shown in Fig. 7, is a 25 × 51 × 25-mm (1 × 2 × 1-in.) rectangular-box-shaped piece with a 1.27-mm (0.050-in.) wall and a semicircular cutout of 12.5-mm (0.5-in.) radius on one side. There are two ribs on one end of the part, leaving details less than 6 mm (0.25 in.) wide but more than 18 mm (0.75 in.) deep to electroform. This is difficult geometry for the electroforming process. However, closely spaced ribs are commonly found on plastic parts. The two-sided SL electroforming mandrel had a parting line thickness of 12.5 mm (0.5 in.) The part was edge gated so the runner and gate were built on the parting line.

Field control devices were mounted over the part prior to placing it in the electroforming tank. These helped create a more balanced current density around the part surface. After about 1 week of plating, the core side had nickel thicknesses that ranged from 0.6 mm in the deepest recesses of the pattern to

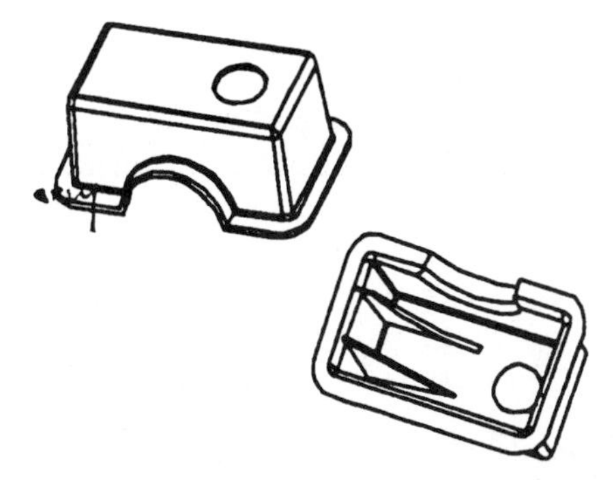

**Figure 7** Kodak test part geometry. Size is 25 × 51 × 25 mm.

more than 2.5 mm at the parting line. The cavity side had even greater nickel thickness plated over the highest part of the model. This is fortuitous because the cavity of the tool sees the highest tensile stresses when parts are molded.

A bonding layer was applied to the back of the nickel shell and to the mold frame. Normally, conformal cooling lines would be placed between the shell and the steel frame, but given the size of the tool, it was decided to simply rely on conduction of heat through the ceramic to the mold frame [a master unit die (MUD) insert]. The buss bar clamp frame was then bolted to the machined MUD base and the nickel shells were backed in sequential fashion then cured hydrothermally following the procedures outlined in the previous section. Some machining was done on the buss bar clamp ring because it did not sit flush on the MUD frame. The finished mold halves shown in Fig. 8 were then returned to Kodak for measuring and part molding.

Preliminary measurements prior to shipment, subsequently confirmed by Kodak, showed that the parting line had distorted a small amount. The core side was concave by 0.1 mm (0.004 in.) and the cavity side was convex by 0.2 mm (0.008 in.) This distortion is believed to be due to the restraint of the plastic model at its perimeter by the buss bar clamping frame, coupled with the differential thermal expansion of the metal and plastic that occurs at the nickel electroforming temperature. Overall, it was observed that there was a slight expansion of the core and cavity geometry relative to the SL mandrel by 0.03–0.1 mm (0.001–0.004 in.).

After the ejector pin holes were drilled for the knockout and the runner was cut from the edge of the shell to the sprue, the tool was mounted in a press and parts were molded from polystyrene using an injection pressure of 700 bar (10,300 psi). Steady-state cavity and core surface temperatures of

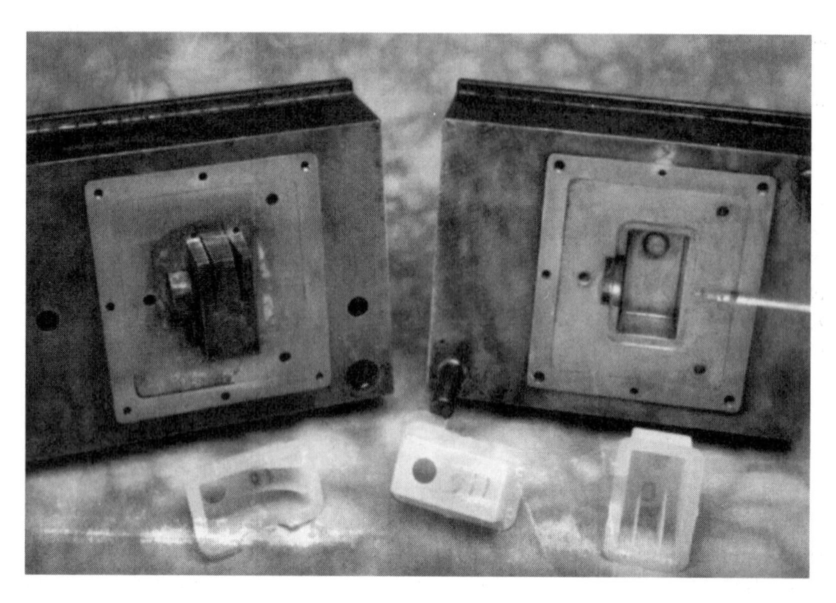

**Figure 8** Kodak tool core and cavity NCC inserts in MUD frame. Molded parts are shown in foreground.

50°C (122°F) and 60°C (140°F), respectively, were measured using a molding cycle of 40 s. (A steel tool for the same part ran on a 30-s cycle.) This was considered quite reasonable given the fact that no special cooling provisions were included in the mold. In order to get the tool to run on a fully automatic cycle, injection pressure was reduced to 380 bar (5600 psi). A total of 5000 high-impact polystyrene parts were molded. The tool did not show any wear or deterioration from running these parts.

After injection molding, the parts were sectioned and the wall thicknesses were measured. They ranged from 1.2 to 1.5 mm (0.048 to 0.060 in.). Nominal wall thickness was 1.3 mm (0.052 in.). The largest differences were seen between the side with the gate and the opposing side that had the 0.5 in. radius cut out. The gated side was thickest. It is not known if this is due to differential pressure from one side relative to the other, or a slight shifting of the core on assembly. The earliest parts run at a pressure of more than 700 bar had a small amount of flash in one corner, but this was eliminated when the molding parameters were optimized for automatic operation.

When Kodak completed their durability test, the mold was returned to CEMCOM Corp. so that it could be used for longer-duration durability trials.

**Table 2** Durability Trial Run on Kodak Test Tool

| Material | Cycle time | No. of parts |
|---|---|---|
| Polystyrene | 40 s[a] | 5,000+ |
| Polyethylene | 30 s | 15,000+ |
| PVC | 30 s | 5,000+ |
| Polypropylene | 30 s | 5,000+ |
| PBT 30% glass | 22 s | 5,000+ |
| Nylon, 30% glass and mineral | 15 s | 5,000+ |
| ABS 30% glass | 25 s | 5,000+ |
| Polycarbonate | 30 s | 1,000+ |
| Total no. of parts molded | | 46,000+ |

[a] Run by Kodak; no cooling in mold.

Prior to running these tests however, cooling lines were added to the mold frame and the surfaces were dressed so that better release would be obtained. This allowed the tool to run fully automatic on a faster molding cycle. Table 2, shows the materials, the cycle times, and the volume of parts molded from each resin. As can be seen, the NCC mold handled both corrosive [poly(vinyl chloride) (PVC)] and abrasive, glass-reinforced materials well. Electroformed nickel is very corrosion resistant, so the performance with PVC was as expected. However, the nickel surface is softer than most tool steels. Whereas molding 15,000 parts with glass-reinforced resin did not change any dimensions on the inserts, even around the gate, there was a noticeable polishing of the nickel surface in this area. Sticking of polycarbonate to the core surface ultimately caused the tool to fail, but more than 46,000 shots were run. This trial demonstrates that the NCC system has the capability to injection mold reasonable production volumes.

## VII. A LARGE NCC TOOL

The performance of the NCC tooling process, when integrated with the capabilities of the CAD and rapid-prototyping systems described throughout this book, are important steps in proving the viability of the process. However, given that the most compelling market for this technology is larger molds, the intent was to prove that the method is suitable when scaled up. This portion of the development effort was performed in conjunction with Pitney Bowes, who needed a large internal part for one of their mailing machines. Figure 9 shows three views of SL tool mandrel for a part that is 385 × 125 × 85 mm deep. With the addition of the extra material for the stepped parting line and the clamping frame around the perimeter, a mandrel 480 × 230 × 100 mm (19 × 9 × 4 in.) was made. The reader can see in these three views the detail for the ejector side geometry on the left, the injector side detail on the right, and the alignment and registration of these detail in the center view. A few of the features on this part were narrow slots which would be difficult to electroform. These small features were machined from metal and fitted to the mandrel prior to electroforming. They were then simply captured in the nickel shell as it was formed. In addition, due to the large size of the part and the need for high precision in the boss locations, it was decided that these would be drilled and placed in the tool after the shell had been formed and captured in the mold frame.

Figure 10 shows the NCC tool after demolding but prior to final machining for the knockout system, pins, and the sprue bushing, which was located

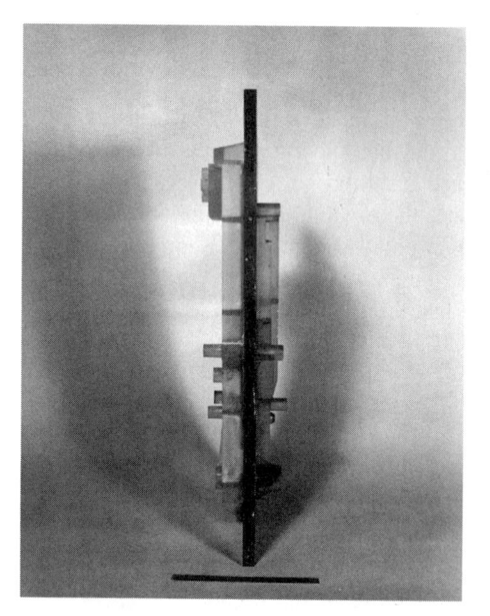

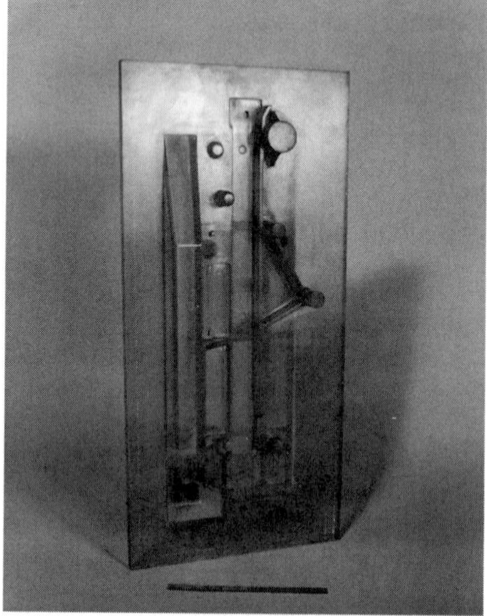

**Figure 9** Three views of the SL mandrel used to make mailing machine part for Pitney Bowes.

**Figure 10** Nickel ceramic composite tool prior to final machining of the knockout system. The ruler in front of the tool is 600 mm (24 in.) long.

near the center of the part in this case. Note the cooling lines emerging from the left and right sides of the tool. These were cast into the ceramic. The fabrication time from receipt of model to the point shown in this photograph was just under 5 weeks. The final machining and finishing brought the total to six *weeks*. Some fitting and bluing was required, particularly on the inclined shutoff areas where a good deal of ''stair-stepping'' had to be manually removed from the SL mandrel. This left the nickel just slightly proud in these regions.

Parts were molded on the tool using a 30% glass-fiber-reinforced Noryl PPO structural foam thermoplastic. Molding cycle was nearly 3.5 min., which is ~15% longer than the part would have run in steel. A photograph of the front and back of the molded part is shown in Fig. 11. The part on the right is turned to show the injection side. The white mark where the sprue has been removed is visible. Perhaps the most notable feature of this large injection-molded part is the fact there is no flash. The part came out of the tool very cleanly, demonstrating the viability of the two-sided mandrel approach for injection molding larger-sized parts.

The part was measured as a quality control check prior to fitting to a functional mailing machine. The holes in the bosses, that must line up with

**Figure 11** Two parts showing the ejector side and injector side detail of the internal part for a Pitney Bowes mailing machine. The ruler in the foreground is 30 mm (12 in.) long.

other features, were within 0.1 mm of their intended location. This is no surprise because these pins were fitted to the tool *after* the forming process. The overall length of the part was nearly 0.6 mm longer than the design length. Given the thermal expansion coefficient of the SL mandrel's epoxy photopolymer, this dimensional change is consistent with the expected thermal expansion of the mandrel at the electroforming temperature. In the future, a simple thermal expansion correction of 0.13% of the mandrel dimentions during the RP&M step should bring the size of the part within a still tighter range. Even so, the part as made was *within specifications*. It fit and functioned well with the other components in the system.

## VIII. AN APPEARANCE PART MOLD WITH COMPLEX FUNCTIONALITY

Having demonstrated that the NCC was suitable for making a large mold, an appearance part mold was fabricated, again with the assistance of Pitney Bowes. One additional ground rule for this tool was that it had to make production quality parts with standard production tool functionality. This meant that a finished part, requiring no trimming or secondary machining, had to be produced by the tool. In addition, the tool had to function automatically so that parts could be molded without an operator standing by the press. The part used to demonstrate this capability is shown in the right side of Fig. 12 along with the "single integrated matched plate electroforming" (SIMPLE) SL mandrel used to create the tool geometry.

The overall part size is 12 in. × 6.5 in. It is small enough to be edge gated, but it was set up with a hot sprue and center gated because larger parts would require this configuration. To avoid a mark on the appearance side of the part, the hot sprue was located on the ejector side. Also note that there are hinges on the long straight side of the part which have a snap fit for mounting. These hinges have a tall core in their center, which could not be electroformed. Rather, this feature was made from a machined metal piece and mounted to the mandrel so it could be captured in the nickel when the model was electroformed. The snap-fit feature, which required a slide action, was formed in the shell by drilling a hole in the insert and inserting a pin through this hole and the snap-fit detail. This formed the hole in the nickel shell through which the slide action moved. Also note that the angle of the hinge features in the tool made it necessary to step the parting line on three sides in order to get a straight pull of the part from the mold.

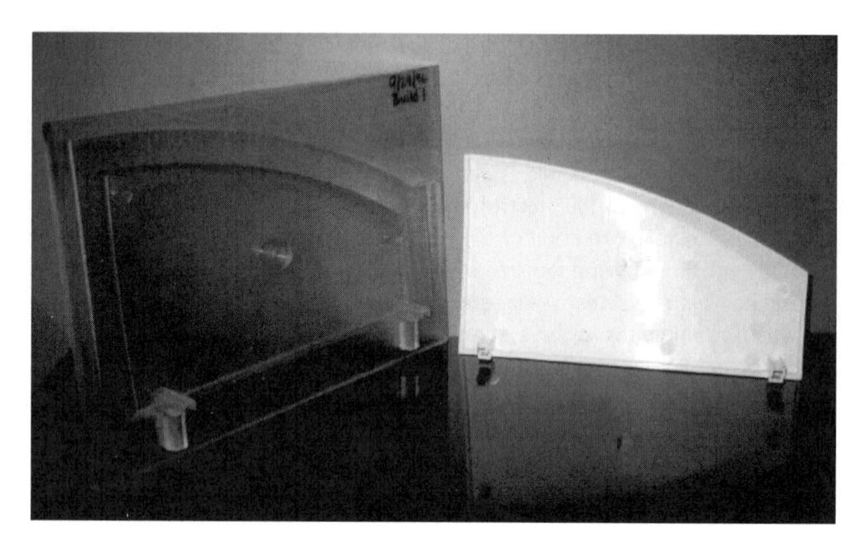

**Figure 12** A SIMPLE tool SL mandrel of appearance part (left), shown opposite a molded part (right) to show relationship. (SL mandrel supplied by 3D Systems.)

Because this was an appearance part which had a texture on the surface, there was some discussion about applying texture directly to the model at the beginning versus texturing the nickel surface after the tool had been sampled. (The latter is standard practice in steel tools.) As it would be extremely difficult to repair any flaws in any textured area while assuring that this repair would remain invisible in the molded part, it was decided that the appearance side of the tool would be made smooth, and textured later. There was some concern that texturing the nickel may be difficult due to the metal's corrosion resistance. However, Moldtech was able to apply the desired texture using a standard mask with a stronger than normal etching solution.

Because tool shutoffs are built into the SIMPLE tool model, there is high accuracy required in the region just outside the perimeter of the part. Tolerances here need to be held to within 0.002 in. This kind of accuracy in the build direction is difficult to achieve with the existing RP&M processes. When coupled with a build layer thickness of 6 mils, the shutoff areas needed a significant amount of hand work on both the mandrel and the nickel shell. Note also that in the hinge area, the parting surface steps down to the level of the detail resulting in a very steep shutoff region.

Once the SIMPLE tool mandrel had been carefully sanded down to at least a 600-grit finish, the features to be captured in the shell were mounted

**Figure 13** Appearance part tool in press. The textured side of the mold is on the left. Reverse ejection with slide action features are shown on the right.

to it, then coated with a conductive layer, placed in the plating tank, and attached to a power rectifier. The mandrel was plated for 10 days and, in this time, nickel was built up to an average thickness of 0.150 in. Inside corners and recessed areas had significantly less nickel than outside corners, but the ceramic backing fills in the unevenness so that the shell will be uniformly supported in operation. The shells produced remained on the model while a proprietary bonding layer was applied. Cooling lines were also mounted three tube diameters behind the shell and special attention was directed to areas of the core which were expected to be the hottest. The SL mandrel was then used to locate the core and cavity in the mold frame, and the CBC backing was applied. After the ceramic set for at least 24 h at room temperature, the mandrel was removed and the NCC tool was postcured.

The core side of the NCC mold was then machined to accept the hot sprue bushing, and holes were drilled through the nickel and the ceramic for each of the ejector pins. The mechanical pulls for the slide actions were also mounted to the frame at this time, in a fashion similar to a machined metal tool except that they were aligned to the preformed holes in the shell and hinge core inserts. Preliminary finishing was done on the nickel at this time, which included quick bluing of the parting plane. The tool was then sent to an injection-molding house to be sampled. The tool is seen in the molding press in Fig. 13.

Minor flashing was found at the corners, and the hinge detail was difficult to fill without burning. Also, the shallow draft hinge area tended to hang up in the tool as parts were run. To overcome these problems, the tool was more thoroughly blued to obtain proper closure around the entire perimeter. Some buildup was necessary and this was accomplished via microwelding. Once the closure was correct, the cavity side was textured and the core detail was more carefully finished around the hinges. Small ejector pins were added to the lowest part of the hinge-forming area in the tool (primarily for venting), and extra ejector pins were added adjacent to each of the hinges. With these modifications, the tool produced the parts shown in Fig. 14.

## IX. COMPRESSION TOOLING

The tool model designs discussed are well suited to simple closures of matched die molds. There is significant application for this technology in the compres-

**Figure 14** Molded parts from the NCC tool. The textured side is on the left and the injection/ejection side is on the right.

sion-tooling market where full positive closures are the norm. Expanding a tool mandrel in the mold opening direction for this type of closure would require that the tool model be substantially thicker and therefore more costly. Additionally, the closure on the core side of these tools are more likely to be eroded or degraded and it would be preferable that these elements be made from a harder material than Ni. To overcome these problems, a method to build aligned shear closures by mounting the mandrel on a machined steel insert was developed. The insert ultimately becomes the core side shear edge. This is shown in Fig. 15. Another feature in this tool is a welded plate mold box rather than a standard mold frame. This can be done because low-volume compression tools can often be run with poppettes rather than a full ejector plate with pins. Also, compression tools are often aligned at the center of the tool using heel blocks rather than pins placed in the corners.

This low-cost tool fabrication approach has been demonstrated on two tools, in conjunction with Zehrco Plastics. The second of these, a two-cavity mold for a sensor case, was made in 3.5 weeks from reciept of CAD data and ran at standard operating conditions for a polyester bulk mold compound of

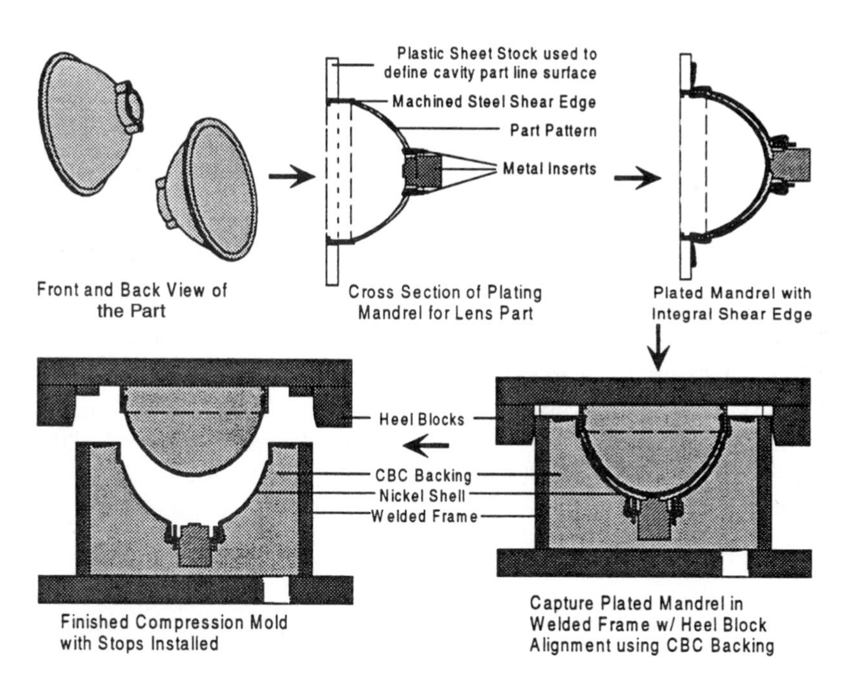


**Figure 15** Fabrication sequence for full positive closure tools such as compression molds.

325F and over 1000 psi. The mold produced more than 1500 parts. Molds using this design should work just as well for sheet molding compound. This basic approach has also been used for tools to mold rubber.

## X. CONCLUSIONS

Nickel ceramic composite toolmaking involves the integration of materials forming and CAD in a process that is consistent with the stability of RP models. It very accurately reproduces the geometry of a forming mandrel so it can produce accurate and aligned tool surfaces without reliance on substantial fitting or finishing work, although the level of finishing required is very dependent on the quality of the models produced from the RP machine. The inherent precision of electroforming also makes it possible to easily fit components made by more traditional methods and simplifies scaling of the process as was demonstrated by the case studies discussed. The RP mandrel making and electroforming processes are both unattended batch processes, and consequently, are not labor intensive. By combining electroformed nickel shells with standard mold frames and a high-strength ceramic backing, a modular assembly is made that can be fabricated rapidly and economically.

The durability limit of the NCC tools is currently under investigation, but a small tool has run more than 45,000 parts. Also, a large, complex injection mold done in conjunction with Pitney Bowes, involving slide actions and a heated sprue, has produced more than 10,000 parts. In these durability tests, the tools were run under standard operating conditions used for machined metal injection or compression molds and they have successfully run both filled and unfilled engineering thermoplastic and thermoset resins. Tools have been made which show that (a) the NCC surface can be textured, (b) that standard metal-hardening methods can be applied to the electroformed nickel surface and (c) inserts can be used where geometric features are not well suited to the electroforming process. The combination of rapid, low-cost forming in a durable tooling system make the NCC system an attractive rapid mold-making process, particularly for larger parts.

## XI. FUTURE WORK

Having demonstrated that the NCC is a viable mold-making process, the practical limits need to be explored for incorporation of slides and lifters. Their incorporation will be tested in different stages of the NCC tool fabrication process to see how they affect the overall fabrication costs and timing. The impact of the design features outlined in the chapter will be assessed on the overall fabrication time of the NCC tooling. It is imperative that methods and process sequences developed do not extened the lead times possible with the NCC tooling method. To this end, more effort will be directed at standardization and modularization so that the process speed can be further improved. Finally, the part size envelope will also be explored. Components that are nearly 0.5 m × 1 m are under consideration for fabrication from RP models. The process is suitable for even larger parts, but it may not be feasible to use current RP model-making methods.

## ACKNOWLEDGMENTS

The author gratefully acknowledges the contributions of the following people and organizations to this effort: Doug Van Putte and his associates at Kodak for providing the SL mandrels, the MUD frame, and for molding parts in the small demonstration tool; Vadan Nagarsheth and Glen Randmer of Pitney Bowes Inc. for continued support of the NCC tooling effort and their contributions to both part design, tool design, and CAD file preparation for the mailing machine part; Mike Naylon of QMI for work with the phone insert part; and Rob Tanis, Lee Robinson, and Kevan Jones of CEMCOM Corp. for their NCC tool-fabrication efforts.

## REFERENCES

1. W Durden. A successful team approach to rapid tooling. SME Rapid Prototyping and Manufacturing Conference Proceedings, 1996. T Gornet. Experiences with DTM RapidTool. SME Rapid Prototyping and Manufacturing Conference Proceedings, 1996. F Prioleau. Comparison of SLS RapidTool process to others. SME Rapid Prototyping and Manufacturing Conference Proceedings, 1996. E Sachs. Injection molding tooling by three dimensional printing. SME Rapid Prototyping and Manufacturing Conference Proceedings, 1996.
2. KR Denton. Quick Cast and rapid tooling: A case history at Ford Motor Company. SME Rapid Prototyping and Manufacturing Conference Proceedings, 1994. R Erikson. Cast tool prototyping for injection molding: Where is it going? Sixth International Conference on Emerging Technologies and Business Trends in Plastics Injection Molding, March 1996.
3. LE Weiss, EL Gursoz, FB Prinz, PS Fussel, S Mahalingam, EP Patrick. A rapid manufacturing system based on stereolithography and thermal spraying. Manuf

Rev 3(1):40, 1990. K Maley. Using stereolithography to produce production injection molds. ANTEC '94, p. 3568. C Hefright. Applying laser technology to rapid prototyping. ANTEC '93, p. 406. A Mathews. Nickel vapor deposition tooling for the plastics industry. Proceedings of the Third International Conference on Advances in Polymer Processing, March 1993.
4. Modern Plastics Encyclopedia '94. New York McGraw-Hill, 1994, p. 185.
5. Modern Plastics Encyclopedia '94. New York: McGraw-Hill, 1994.
6. L Miller, S Wise. Chemically bonded ceramic tooling for advanced composites. Mater Manuf Process 5(2):229–252, 1990.
7. S Wise. Net shape nickel ceramic composite tooling from RP models, SME Rapid Prototyping and Manufacturing Conference Proceedings, 1996.
8. DA Van Putte, LE Andre. A step-by-step evaluation of building an investment cast plastic injection mold. SME Rapid Prototyping and Manufacturing Conference Proceedings, 1995.

# 7

# Nickel Vapor Deposition Technology

**Debbie Davy**
*Mirotech, Inc.*
*Toronto, Ontario, Canada*

## I. WHAT IS NICKEL VAPOR DEPOSITION?

Nickel carbonyl vapor deposition (NVD) is a novel metal-forming process based on the growth of a metal from gaseous vapors, and it has evolved from what was once a refining process into a method for quickly making extremely accurate thin-shell molds. The basic chemical reaction is given by

$$Ni + 4CO \xleftrightarrow{110\text{-}190^{\circ}C} Ni(CO)_4$$

These molds can be used in conjunction with rapid-prototyping patterns in many diverse applications, such as injection molding, blow molding, net nickel shapes, and so forth. This technology allows mold-makers, molders, original equipment manufacturers, and others to respond to rapidly changing markets, reducing costs and shortening product-development time. The NVD process offers the potential for rapid turnaround times, accuracy on an atomic level, and faster cycle times relative to existing manufacturing processes.

Figure 1 is a schematic of the NVD process, utilizing nickel carbonyl gas, $Ni(CO)_4$. This gas provides a useful property: Specifically, it breaks down into a solid metal at a high rate of decomposition. By heating a mandrel or substrate to the required temperature and having $Ni(CO)_4$ gas flow over the mandrel in a sealed chamber, an exact nickel negative of the mandrel is obtained. Deposition rates from 50 to 750 μm/h (0.010–0.030 in./h) can be obtained on surfaces held at temperatures between 110°C and 190°C. The nickel

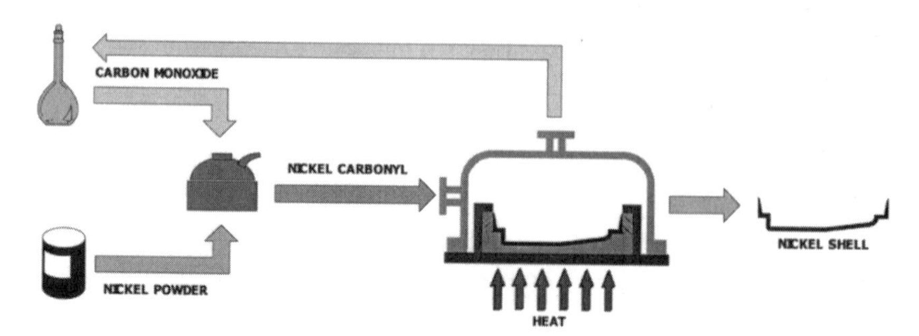


**Figure 1** Schematic of the NVD process.

dendritic crystal size can be controlled by changing the process parameters. NVD's unique features are recognized by the scientific and industrial community for their ability to reproduce the surface detail of a master down to the finest texture and smoothness without the residual stresses and warpage often found in other deposition processes.

In the NVD process, as in any other chemical vapor deposition (CVD) process, surface preparation and cleanliness are critical. State-of-the-art surface cleanliness inspection equipment must be utilized to ensure repeatable high-quality deposits in an industrial environment.

## II. THE NEED FOR NVD

Over the past 10 years, all major global manufacturers have experienced tremendous competitive pressures. The Japanese and Pacific Rim countries have utilized their low-cost, high-quality production techniques to gain a significant level of penetration into both North American and European markets. These competitive pressures are forcing North American and European manufacturers to look for new innovative technologies to help them in reducing the price of their product and especially in reducing their product-development time.

There is, however, a major problem associated with traditional moldmaking techniques which results in significant delays from the approval of the part design to the completion of the finished tool. This is the time required to machine the core and cavity inserts of a mold to the precise dimensions required, typically resulting in mold delivery times from 16 to 30 weeks.

Nickel vapor deposition technology is an ideal vehicle for these new trends, as it can have a dramatic impact on product-development time, typically providing a tool in less than 6 weeks at a reduced cost relative to conventional computer numerically controlled (CNC), machining or electrical discharge machining (EDM). NVD technology potentially has numerous applications across a broad range of industries, applying to virtually all molding technologies in use today.

## III. A BRIEF HISTORY OF CVD PROCESSES

The deposition process via nickel carbonyl gas was first discovered in the late 19th century by Ludwig Mond in Wales, U.K. His company was a predecessor of International Nickel (INCO). The process was used to refine nickel and also to make nickel powders and pellets on a large industrial scale.

In the 1950s, the U.K. and U.S. nuclear industry first recognized the attractiveness of the carbonyl deposition process where plutonium pellets were encapsulated with nickel in weapons research. Naturally, this application was secret and, hence, did not develop commercially.

In the 1960s, in the United States, a small private company commenced research into practical commercial applications using nickel carbonyl deposition with a chemical process simplified from the original Mond process. Several patents ensued, but no significant industrial interest resulted. A commercial entity, Vapourform Products, was formed and changed hands several times. Eventually, it became a division of Detroit-based Formative Products Inc., which went out of business in early 1990.

In the early 1970s, in Kitchener, Ontario, a company called Spraymold was formed to develop a particularly novel molding process requiring highly specialized tooling. In parallel with their molding process, the company assembled a simple, functional nickel carbonyl deposition facility to manufacture small nickel tools. Nonetheless, Spraymold also closed in the late 1970s.

Today, Mirotech Inc. (Toronto, Ontario, Canada), Mirotech's licensee Galvanoform GmbH (Lahr, Germany), and Mirotech's former partner INCO (Copper Cliff, Ontario, Canada), currently make nickel *shells* via the chemical vapor deposition of nickel. There are several small U.S.-based enterprises that are involved in the nickel coating of powders and fibers via the carbonyl process. There are also large commercial plants in the United States and Europe that use iron carbonyl (similar chemistry to nickel carbonyl) to manufacture the iron powder used in magnetic media.

## IV. ADVANTAGES OF THE NVD PROCESS

1. Shapes/cavities are created with uniform wall thickness throughout. External corners build up at the same rate as flat surfaces.
2. Internal corners do not show a significant reduction in deposit thickness, resulting in a uniformly strong tool face.
3. Deposition can be achieved on appropriate substrates without affecting the surface fidelity of the master.
4. Nickel provides a mold surface with high-temperature and corrosion-resistant properties suitable for molding aggressive resins such as phenolics or vinyls.
5. Nickel offers excellent release properties.
6. Nickel vapor deposition produces parts with low levels of internal stress. This results in minimal warping of the finished shells, better matching cores, and cavities.
7. The NVD process is generally less expensive than traditional tooling, offers fast turnaround time and rapid deposition (e.g., a 12.7-mm-thick NVD nickel shell can be generated in less than 2 days, regardless of the tool size). The NVD process typically deposits nickel at rates from 50 to 750 μm/h (0.010–0.030 in./h).
8. The surface of the master is replicated with outstanding surface definition; for example, optical quality surfaces have been replicated.
9. Uniform shell wall thickness is produced, even around acute angles. The NVD process relies on the *thermal* decomposition of nickel onto the substrate and not an electrolytic deposition. This means that if the surface temperature of the substrate/mandrel is uniform, the thickness of the deposit will also be uniform! This is true regardless of the geometry or shape of the substrate surface. Sharp internal corners can be used, resulting in increased design freedom.
10. The NVD nickel molds can be repaired by TIG welding or brazing.
11. The NVD nickel is more dense and ductile than electroformed nickel. This results in a tougher and stronger tool face capable of withstanding additional repeated tooling process cycles and a reduction in the tendency for the nickel tool face to crack in use.
12. The NVD nickel has very little porosity. This provides the following benefits: (a) NVD nickel polishes extremely well; (b) there is no fear of breaking through the surface into porous areas of nickel; and (c) NVD nickel contains almost no sulfur whatsoever and can

be welded readily. This means that repairs or modifications to the NVD tool surface can be performed at any time, as the hard surface layer produced in the NVD process is not technically an alloy. Only the microstructure of the nickel is changed in the hardening process. NVD nickel offers tool faces with a hard abrasion-resistant active surface.

13. The NVD process parameters can be changed at the start of the deposition cycle to produce a hard surface. When a desired hard-skin thickness has been reached, the process parameters can be *gradually* changed to normal deposit hardness, eliminating an abrupt interface which could cause delamination. Furthermore, NVD is an *in situ nickel deposition technique*, which eliminates the need for mandrel metalization.

## V. DISADVANTAGES OF THE NVD PROCESS

1. Stereolithography mandrels cannot be used at the required NVD temperatures.
2. ACTUA mandrels cannot be used at the required NVD temperatures.
3. Sanders models cannot be used at the required NVD temperatures.
4. Selective laser sintering (SLS) wax models cannot be used at the required NVD temperatures.
5. Nickel carbonyl and carbon monoxide are very toxic, requiring special handling procedures and facilities.
6. There are only a few sources for the process; hence, potential delays could result due to job queuing.

The NVD process may not be advantageous in those applications requiring small parts that can otherwise be made by very simple machining techniques.

Also, in some rapid prototyping and manufacturing (RP&M) applications, the use of NVD may not be appropriate, as it delivers a precise replication of the pattern. When the mandrel comes from one of the RP&M technologies, stair-stepping imperfections will be more evident with the exact replication of the NVD process. Cleaning, sanding, and polishing of the RP&M mandrel becomes critical.

However, the greatest advantage of NVD in rapid tooling applications is that the resulting NVD shell can be used for final part production. NVD's ability to quickly deliver a precise replication of a master is also an advantage in getting products to market more quickly.

## VI. NVD APPLICATIONS

The NVD process has been used in many diverse applications, such as the following:

1. Injection molds: optical applications and gears
2. Compression molds: battery plates
3. Blow molds: engineering resins and large tools
4. Resin transfer molding, resin injection molding, sheet molding compound, slush molding, and autoclave molds
5. Nickel tooling: gloves, compact disk stampings, pulp trays, and foundry patterns
6. Nickel shapes: specialty tubing, laser mirrors, diffraction gratings, erosion shields, wave guides, EMI shielding, structural panels, and pressure vessels
7. Stamping dies

## VII. PROPERTIES OF NVD NICKEL

The work on nickel that has been deposited by CVD has yielded data that falls into several categories. These are tensile properties, hardness properties, thermal expansion, residual stresses, chemical composition, and microstructure. These data have been obtained from tests and experiments that have been performed at Ortech, at the University of Toronto, at Ontario Hydro's Research Division, and at Camnet.

Coefficient of thermal expansion: $13.1 \times 10^{-6}$ m/m/°C
Thermal diffusivity: 3.66 $m^2/h$
Thermal conductivity: 88 W/m°C
Residual stress (surface): 30–60 MPa tensile
Yield strength: 584 ± 39 MPa
Ultimate tensile stress: 827 ± 7 MPa
Modulus elasticity: 157–224 GPa
Elongation: 6–12.4%
Hardness: 15–50 RC (variable)

| Chemical composition: | Ni | 99.98% |
|---|---|---|
| | C | < 150 ppm |
| | S | < 1.0 ppm |
| | H | < 7.3 ppm |

Metallographic work using a transmission electron microscope found that NVD nickel possessed a dendritic structure, as shown in Fig. 2. The samples were taken from material at various depths in the as-deposited sheets. The grains were found to fall into two size ranges. The large grains were separated by many smaller grains, which were equiaxed. Many large grains were found to be twinned. Most large grains had a low dislocation density and no significant strain. It was also found that the average grain size increased with distance from the substrate. Near the substrate, the microstructure consisted mainly of fine grains about 0.1 μm in size. Various reports suggest that the preferred growth of these fine grains in certain crystallographic directions accounts for the dendritic-type microstructure exhibited by the as-deposited NVD nickel.

Although it appears that the fine grains do not show any texture, the larger grains do. The twin boundaries in the larger grains have a preferred orientation that suggests the ⟨110⟩ direction in these grains lies in the growth direction of the substrate.

Figure 3 shows an example of how it is possible to control the crystal size of the deposited nickel. By changing the deposition parameters, the initial grain size was decreased in the middle range of the sample.

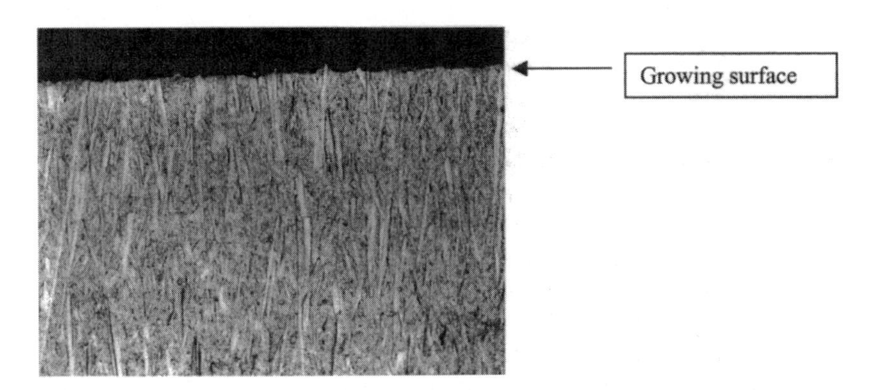


**Figure 2** Typical grain structure for CVD nickel.

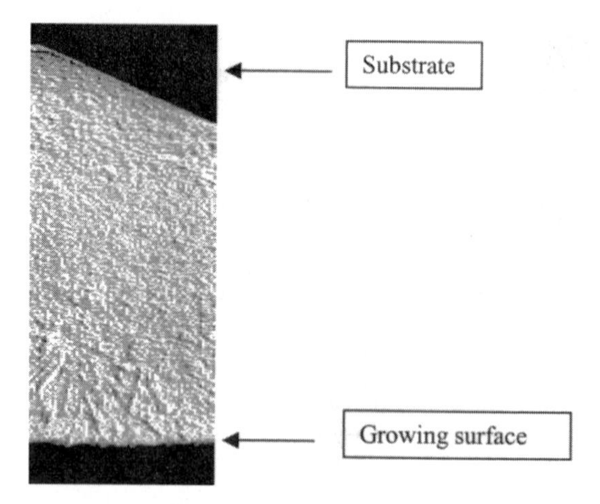


**Figure 3** Example of control of crystal size during the deposition of nickel.

## VIII. COMPARISON BETWEEN NVD AND ELECTROFORMED NICKEL TOOLING

Both NVD tooling and electroformed tooling can offer substantial cost-saving benefits over machined steel tooling, especially for larger parts. Also, a substantial time savings can often be realized relating to shorter tool-generation times. Electroformed nickel, although also capable of generating production tooling, has some inherent limitations that must be considered before a successful tool can be designed.

The NVD process relies on the *thermal* deposition of nickel onto the substrate, not an electrolytic deposition. This means that provided the surface temperature of the substrate/mandrel is uniform, the thickness of the deposit will also be uniform. This is true regardless of the geometry or shape of the substrate surface. With electroformed nickel tooling, surface geometry has a marked effect on deposit thickness. External corners on the mandrel exhibit higher local electric fields and will show an increased buildup of nickel. Conversely, internal corners result in lower local electric fields and will show a decrease in deposit buildup, as seen in Fig. 4A. Very sharp internal corners will accentuate this effect. The result can often be an internal corner with 50% or less nickel deposit than the average thickness. The thinner nickel shell will be inherently weak, which can result in premature failure of the tool in this location. This inherent electroforming limitation often results in a redesign of the mandrel to minimize shell thickness variations.

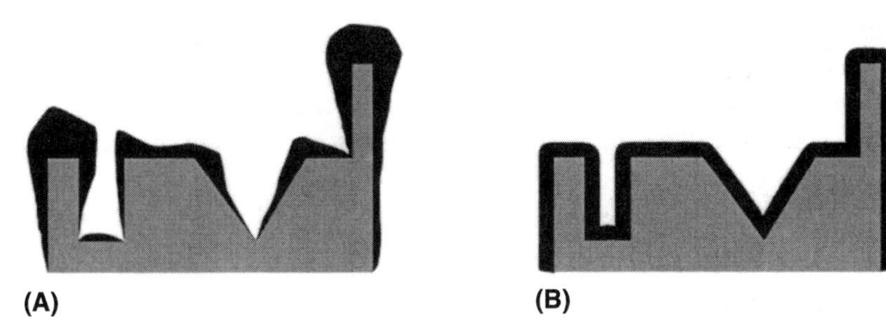


**Figure 4** (A) Electroformed nickel; (B) NVD nickel.

With the NVD process, external corners build up at the same rate as the balance of the surface, as shown in Fig. 4B. The internal corners do not show a significant reduction in deposit thickness, resulting in a much more uniform shell thickness and, hence, more uniform shell strength. Very sharp internal corners can be used, providing increased design freedom. With electroforming, this is not always possible.

The NVD nickel is considerably denser and much more ductile than electroformed nickel. This results in a tougher and stronger tool face capable of withstanding a greater number of repeated tooling process cycles, with a marked reduction in any tendency for the nickel tool face to crack in use. Electroformed nickel often contains both surface and internal porosity. NVD nickel has very little porosity.

The NVD nickel polishes better than electroformed nickel, and there is no fear of breaking through the surface into porous regions of nickel. Electroformed nickel can contain sulfur, which results in poor welding performance. NVD nickel contains almost no sulfur whatsoever and can be welded readily. This means that repairs or modifications to the NVD tool surface can be performed at any time.

Both electroformed nickel and NVD nickel can provide tool faces with a hard surface for abrasion resistance. However, when electroformed nickel is produced with a hard face, the process generally requires two steps:

1. A hard skin of nickel alloy (usually nickel/cobalt) is deposited to a nominal thickness (usually 1–2 mm) in a nickel-alloy plating tank.
2. The shell is removed from the alloy plating tank, the rear surface is activated to receive an additional layer of nickel, and the shell is placed in a ''normal'' (pure) nickel plating tank to complete the balance of the required thickness.

A defect that can occur due to this two-step process is *delamination*. If the activation of the rear of the nickel face is not 100% successful, then stress concentrations can occur at the interface between the two different hardness layers. Potential delamination is difficult to inspect and may occur much later during tool operation.

## IX. COMPARISON BETWEEN NVD AND CONVENTIONAL TOOLING

The majority of large molds are made by CNC and EDM machining out of steel. Some tools intended for limited production are also made from aluminum, which is less expensive to machine than steel. For most low-pressure plastic injection-molding applications, steel tooling is overengineering and involves an unnecessary expense. Additionally, many large plastic parts are filled with glass fibers. Hence, abrasion is an important issue to consider in mold design. Consequently, steel often supercedes aluminum as the material of choice. Here, an NVD composite mold can have an enhanced surface hardness to ensure a long-lasting tool surface. A softer substrate such as copper or aluminum-faced epoxy can become a viable alternative to steel molds for moderate to large production runs.

Steel molds are very heavy compared to NVD composite molds. Consequently, handling and storage is an important consideration. Design, fabrication, assembly, checkout, and delivery of a steel mold can take 16–26 weeks or more, whereas an NVD composite mold can be completed in less than half that time.

## X. NVD ENVIRONMENTAL CONSIDERATIONS

Handling of the nickel carbonyl and carbon monoxide requires sophisticated instrumentation and safety procedures, as well as a fully trained staff. Environmental approvals are necessary. Nickel carbonyl and carbon monoxide are very poisonous and require special and careful handling procedures.

An NVD plant does not generate any water, soil, or air pollution, and all waste gas is reclaimed. Solid waste is only in the form of cured resin parts. No liquid waste is generated, and cooling water is not contaminated. The NVD process operates at atmospheric pressure.

## XI. ABOUT NVD MANDRELS

NVD nickel can be deposited onto any surface provided the following hold:

1. It can withstand the operating temperature conditions (up to 190°C) without degradation of accuracy or its mechanical properties.
2. It is reasonably thermally conductive.
3. It does not out-gas during the NVD process.

Unfortunately, this is *not* the case for mandrels made from the following:

- Stereolithography
- ACTUA-2100
- SLS wax
- The Saunders process
- The Cubital solid-ground curing process
- Fused deposition modeled (FDM) waxes (ICW04 and MW01)

The suitability of other RP& M materials must be carefully evaluated. NVD mandrels are typically heated with cast-in heating coils. Occasionally, mandrels are heated on a heated platen (''hot plate'') only if they are highly thermally conductive and relatively flat.

Aluminum, steel, or brass mandrels have been used for the NVD process. Mandrels have also been made by the NVD process itself. This approach permits the fabrication of multiple mandrels from an original. Other materials such as graphite, glass, or ceramics have been successfully used for NVD mandrels in special applications.

A composite Mandrel can be used to make mandrels for the NVD process. A specific composite mandrel approach was developed by Mirotech as a rapid, low-cost, low-labor method of providing a surface suitable to acquire an NVD nickel deposit. The unique feature of this particular NVD composite mandrel is that the *coefficient of thermal expansion* (CTE) has been matched with NVD nickel.

The backing system that has been developed has a matched CTE to the nickel tool face. This *greatly reduces stresses* induced during the injection-molding thermal cycles. The backing system provides an excellent bond to the nickel face. An additional shell attachment aid was specifically developed to ensure an excellent bond between the nickel face and the backing, eliminating any problems with delamination. The process is shown schematically in Fig. 5.

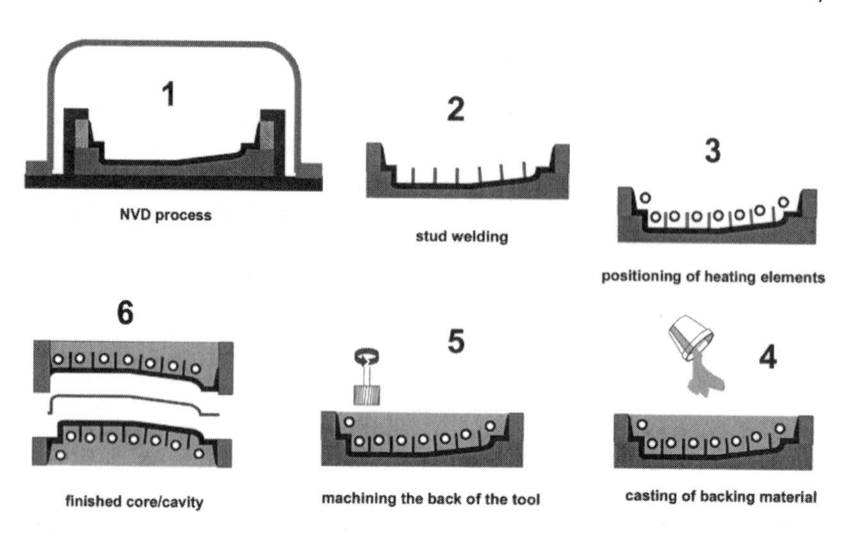


**Figure 5** The production of a composite NVD mold.

Figure 6 shows a cross section of an NVD mold. The metal support frame can be encapsulated *in situ* during the deposition cycle, eliminating postoperations of shell trimming and mechanical fastening of the shell to the support frame. This ensures a stronger attachment of the shell to the frame. The result is that the shell can remain on the mandrel for the casting of the backing system, the mandrel becomes the *support* for the casting of the backing system, and any distortion of the tool is greatly minimized.

Ejector-pin bushings and sprue bushings are encapsulated *during* the NVD process to eliminate the need for a two-step reactivation process; this, in turn, avoids any potential future delamination issues.

Some important features of the nvd composite mold approach are as follows:

- The backing has its CTE matched to that of the NVD shell.
- The backing is well bonded to the NVD nickel shell.
- The ejector-pin and sprue bushings are encapsulated in NVD nickel.
- The steel frame is also encapsulated in NVD nickel.
- The backing material is thermally conductive.

Figure 7 shows how a metal support frame can be encapsulated *in situ* during the deposition cycle, eliminating postoperations involving shell trimming and subsequent mechanical fastening of the shell to the support frame.

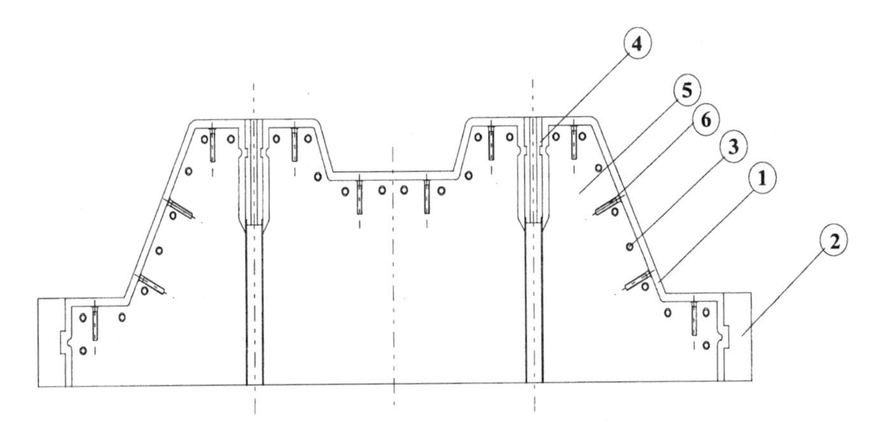


**Figure 6** Cross-section of an NVD mold. (1) NVD nickel face; (2) outer support frame of fabricated steel with encapsulation; (3) heating/cooling coils, usually copper located at rear of shell; (4) ejector-pin bushings and gates (note encapsulation); (5) cast backing (composite with matched CTE); (6) bonding aides (stud welding).

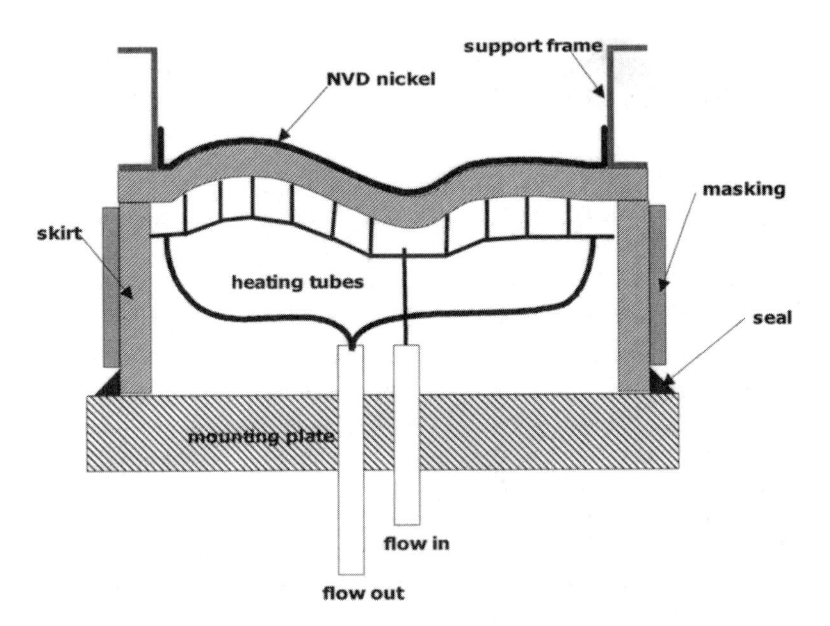


**Figure 7** NVD composite mandrel.

Integral construction also results in a stronger attachment of the shell to the frame. The ability to deposit into sharp internal corners with the NVD process allows this important advantage to be realized. The mandrel becomes the support for the casting of the backing system, resulting in a nearly distortion-free tool. The shell remains on the mandrel throughout the casting of the backing system. This is possible provided the mandrel material has a CTE that is closely matched to that of the nickel tool face.

The need for a composite-type mandrel was recognized early in the development of the NVD technology. Initially, an aluminum-machined mandrel was used in some applications and proved quite successful. However, for large tools and for tools where the starting point is an RP&M model intended to *avoid* expensive machining, the importance of the nonmachined mandrel type is obvious. Indeed, one of the most important features of the NVD process is the fact that CNC machining in steel or aluminum can be eliminated, although there remains a concern regarding the accuracy of RP&M-based mandrels for production-tooling requirements. As an example, in the automotive aftermarket, tooling expense is critical because of relatively short production runs. The NVD composite mold can be made from the supplied plastic part in a shorter time and without the traditional CNC machining process.

## XII. GENERAL DESCRIPTION OF TOOLING FOR THE PLASTICS INDUSTRY

Tooling for the plastics industry is typically produced in two major forms:

1. **Production Tooling**: machined steel; machined aluminum; electroformed nickel shell with cast resin backing
2. **Prototype "Bridge" Tooling**: cast epoxy or fiber-reinforced composites; spray metal shell with cast backing

As their names imply, production tooling is typically used to produce high or moderately high part volumes ($10^3$–$10^6$). Bridge tooling is used for *prototyping* or short-run, low-volume applications (10–$10^3$). Reinforced plastics (composites) such as glass-filled nylon, glass-filled acrylonitrile–butadiene–styrene or glass-filled polycarbonate induce considerable abrasion or wear of the tool and require an abrasion-resistant active tool face.

Finally, NVD tooling can provide high strength, excellent abrasion resistance, good thermal conductivity, and lower internal stresses to maintain accuracy. Coupled with reduced cost and rapid tool generation, NVD tooling is certainly worthy of continued development and evaluation.

## XIII. CASE STUDY

Company X approached Mirotech to make RP&M NVD tooling for a set of can blow-mold inserts. Company X supplied the patterns, made of cast epoxy resin with a heat distortion (HDT) of 420°F. The patterns contained all of the surface textures and details and had a physical size of $4 \times 3 \times 1$ in.

Mirotech also made injection-mold cavity inserts for blow-molding parts in a modular mold base. The inserts were finished to size, including the runner and gate, ejector-pin holes, and water lines. Company X supplied the epoxy part (HDT 420°F) that was used as a pattern.

Mirotech mounted these parts on a tooling plate (with the gate details machined in) and cast epoxy molds to produce NVD mandrels. Mirotech-cast composite NVD mandrels with cast inserts used to locate bushings in the nickel shell for ejector-pin guides. The mandrel supported a steel frame used to create the outer perimeter of the final tool and locating pins and bushings.

Mirotech deposited NVD nickel on both parts at the same time to a thickness of 0.150 in. The support backing (with a matched CTE ceramic/epoxy) was cast *in situ* on the mandrel before removal. The rear face was then machined to the thickness and the shell removed from the master.

From the receipt of the masters, Mirotech delivered the finished inserts to Company X within 2 weeks. The NVD deposition took 1 h. Company X blow molded 20,000 parts from polyethyline. Company X saved 5.5 months in development time and reduced its regular costs by 40% by using NVD nickel. Company X went on to use the prototype NVD nickel shell for the regular production of can blow molds.

# 8
# The ExpressTool Process

**Paul F. Jacobs**
*Laser Fare—Advanced Technology Group*
*Warwick, Rhode Island*

## I. INTRODUCTION

Previous chapters of this book have described the business significance of "rapid time-to-market" and the potential impact of reduced tooling lead time. Also discussed were techniques for the generation of "rapid soft tooling" as well as recent advances in "rapid bridge tooling." Additional chapters describe various approaches to "rapid production tooling," including their advantages and benefits, as well as their shortcomings. In this chapter, we shall describe the ExpressTool™ process in some detail.

ExpressTool evolved from a joint project between the Hasbro Corporation (Pawtucket, RI) and Laser Fare, Inc. (Smithfield, RI). As one of the world's largest toy companies, Hasbro generates a great number of plastic injection molds every year. Aware of the importance of "rapid time-to-market" in the highly competitive toy business, Hasbro formed a strategic partnership with Laser Fare in 1992. During this collaboration, a number of different rapid tooling approaches were investigated. The most successful was based on the electroforming work performed by Richard Barlik, and is the predecessor of the ExpressTool process. Although numerous modifications have been made since, the basic physics and chemistry of the patent pending process were developed jointly by Hasbro Corp. and Laser Fare from 1992 to the present.

When the ExpressTool process had achieved an appropriate level of maturity and repeatability, an extended beta test program was started. This led to the fabrication, assembly, and operation of molds for various corporations that were well aware of the importance of rapid time to market. This test program continued throughout 1997. In March 1997, ExpressTool, Inc. was formed as a wholly owned subsidiary of Infinite Group, which also owns Laser Fare. The ExpressTool process was officially commercialized as of January 1998, for the purpose of fabricating high-productivity *production* tooling. Currently, production molds are being built for automotive, aerospace, and consumer product manufacturers.

## II. HIGH-THERMAL-CONDUCTIVITY MATERIALS

Figure 1 plots thermal conductivity for some relevant mold materials. Heat transferred from the plastic must be *conducted* through the mold before it can be removed by coolant. Thus, the thermal conductivity of the mold directly impacts the speed of the injection-molding process. Inspection of Fig. 8 reveals one of the basic problems with steel, by far the most common material used in building production molds.

Here, H-13 tool steel, having a thermal conductivity of 28 W/m °K was chosen to be representative of the broad class of ''tool steels.'' As a point of interest, 316 stainless steel is even less thermally conductive, at about 20 W/

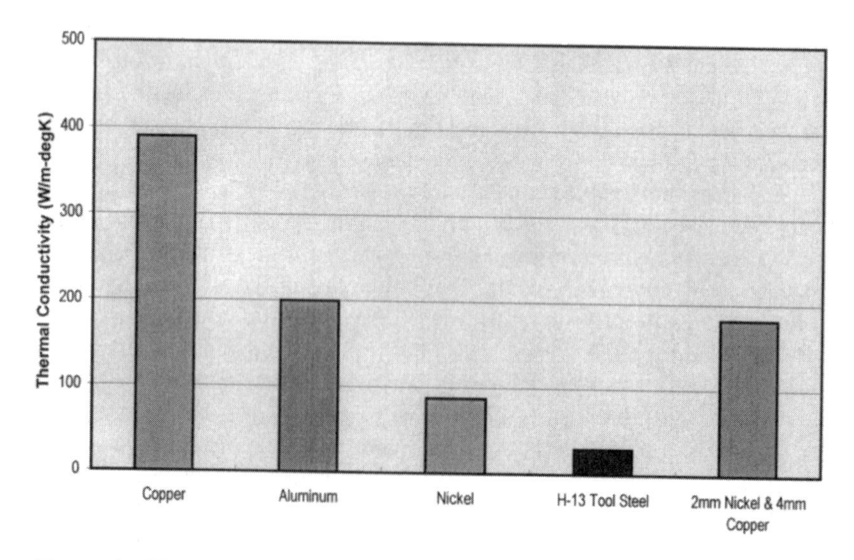


**Figure 1** Thermal conductivity of relevant mold materials.

m °K. By comparison, copper at 390 W/m °K is about 14 times as conductive as H-13 steel and almost 20 times as conductive as 316 stainless steel! Although pure copper is too soft to provide long tool life at the active mold surface, it is a terrific material for mold thermal management.

Next, aluminum possesses roughly half the thermal conductivity of copper, but it is also too soft for long tool life. Nonetheless, as it is easily machined, aluminum is often used for prototype or "bridge" tooling applications, requiring a few hundred to perhaps as many as 50,000 parts injection molded in the desired engineering thermoplastic. If glass-filled plastics are required, aluminum tool life will be further reduced.

On the other hand, nickel has a thermal conductivity of 88 W/m °K; more than *triple* that of H-13 steel and *quadruple* that of 316 stainless steel. Furthermore, nickel is very corrosion resistant, polishes well, is relatively hard (i.e., electroformed nickel has an initial hardness in the range $22 < HR_c <$ 25, but soon work hardens to $30 < HR_c < 35$), is abrasion resistant, can be textured, and provides excellent release characteristics. Combining a 2-mm-thick *nickel shell* at the active mold surface with a 4-mm-thick copper thermal management layer that encapsulates conformal cooling channels (discussion to follow) can provide dramatic benefits. The resulting Ni–Cu composite has an effective thermal conductivity roughly seven times that of conventional steel tools, while capable of generating production part quantities.

## III. CONFORMAL COOLING CHANNELS

Conventional steel tools are generally computer numerically controlled (CNC) machined or electrical discharge machined (EDM) from a solid block of tool steel. Consequently, the cooling channels must also be drilled into solid steel. As a result, these channels essentially consist of a series of interconnecting *straight segments*, each having a *circular cross section*. The drilling operation inevitably results in two important limitations.

First, because the cooling channels are "gun-barrel drilled," they cannot be made to conform to the curved shapes typical of injection-molded plastic parts. The result is that some portions of the plastic are better cooled than other regions. The cooler plastic sectors reach their solidification point earlier than the hotter zones. When the cooler sectors solidify, they shrink. Somewhat later, when the hotter regions finally have cooled sufficiently to solidify, they also shrink.

Unfortunately, the material shrinking last is attached to previously shrunken plastic. This "sector delayed shrinkage," occurring *after* attachment, behaves like a classic bimetallic strip. The result is substantial internal stress and plastic part distortion!

Thus, an important if not often articulated goal in plastic injection molding is to improve the uniformity of the active mold surface temperature distribution over time. Finite-element analysis (FEA) results presented in this chapter show that conformal cooling channels (CCCs), in conjunction with high conductivity mold materials, can provide substantial temperature uniformity benefits. By optimally positioning the CCC in *x*, *y*, and *z* space, it is possible to further reduce mold temperature variance.

A key measure of mold performance is $\Delta T_{max}$, defined as the difference between the highest temperature of the active mold surface and the lowest temperature of the active mold surface at the instant the first sector of plastic begins to solidify and shrink. Lower values of $\Delta T_{max}$ provide more uniform shrinkage and, consequently, less part distortion (1–3).

Figure 2 shows a conformal cooling channel used in the injection molding of a Vaseline jar cap for Chesebrough-Ponds. Note that the CCC transitions from a straight vertical section into an oval shape in the horizontal plane and back to vertical again. Machining a channel of this geometry in a solid block of steel would be impossible in a single piece, or prohibitively complex and expensive in multiple sections. However, when the active surface of the tool has been electroformed as a thin nickel shell, then positioning CCC behind that shell becomes relatively straightforward.

**Figure 2** An example of conformal cooling.

Second, conventional drilled cooling channels (DCCs) have circular cross sections as a natural consequence of the drilling process. From Euclidean geometry, it is well known that of all possible closed two-dimensional shapes, circles have the smallest perimeter for a given cross-sectional area. Coolant flow rate (e.g., gallons per minute) is proportional to the enclosed cross-sectional area.

However, the heat transferred *from the mold into the coolant* is directly proportional to the perimeter of the channel. Thus, a drilled cooling channel with a circular cross section provides the *minimum* heat transfer for a given coolant flow rate. For this reason, a range of CCC cross sections should be explored to determine which shape provides the most effective cooling per unit coolant flow rate. Including such items as boundary-layer effects in the channel, laminar versus turbulent flow effects on the film coefficient, the potential for deflection of the active mold surface/cooling channel under high injection-molding pressures, and the intrinsically asymmetric nature of the heat flow, this is hardly a trivial problem.

## IV. THE EXPRESSTOOL PROCESS

The key aspects of the ExpressTool process are described below. The procedure can best be illustrated by following a typical mold-fabrication sequence:

1. Patterns, or ''mandrels''(the term used when electroforming), are first designed in computer-aided design (CAD). The entire ExpressTool process begins with a CAD model of the mandrel. Most preferable is a three-dimensional (3D) ''solid'' CAD model. Surfaced models are also acceptable, but less favored. ''Wireframe'' CAD models are not appropriate for the ExpressTool process.
2. Mandrels are then fabricated using CNC machining. This approach has been selected for three reasons:

   - Accuracy
   - Dimensional stability
   - Speed

   Major improvements have been achieved in rapid prototyping and manufacturing (RP&M) part accuracy since 1989. Specifically, stereolithography (SL) part accuracy has advanced over the past decade (4). This has been largely the result of (a) improved understanding of the fundamental SL processes (5), (b) photopolymers with reduced shrinkage and improved green strength (6), and (c) the deployment of advanced build techniques (7). Nonetheless, with the possible exception of recent accuracy results by Sanders (which are not yet statistically complete), *none* of the RP&M processes can match CNC accuracy and repeatability for lengths beyond about 4 in. For dimensions over 10 in., RP&M mandrel errors often exceed *three times* the corresponding CNC values.

Further, a number of commercially available, CNC-machined materials currently provide better dimensional stability than any RP&M-generated parts. This is true for ambient conditions (i.e., sitting on a bench, waiting for the next step), and also for electroforming conditions (i.e., being immersed in a vat containing a warm, aqueous, electroforming solution).

Figure 3 shows a CAD model of a mandrel used as a test sample to establish dimensional stability. Initially, this CAD test part was CNC machined in the test material. Next, a number of important dimensional characteristics (e.g., the flatness of both the top and bottom planes, the straightness of the vertical walls, and the values of both the interior and exterior dimensions in the $X$,

**Figure 3** CAD model of a dimensional stability test sample.

$Y$, and $Z$ coordinates) were measured with a coordinate measuring machine (CMM) shortly after machining. Next, CMM measurements were made at the same part locations each day for about 2 weeks. The intent of the tests was to determine how much these measurements change over time.

Figure 4 is a plot of the dimensional stability data obtained for Ciba Express 2000 Aluminum-reinforced polymer board. Here, the difference between the nominal CAD dimension and the measured dimension is plotted as a function of time, for a number of different dimensional features. Measurements of identical features, including the flatness of the top and bottom horizontal planes, as well as the straightness of all machined vertical surfaces, were performed each working day for a period of 16 days.

All test data were gathered with a Brown and Sharpe ''Micro-Val'' Model 454 coordinate measuring machine (CMM). In this format, the trace (i.e., a plot of error versus time) of a perfectly dimensionally stable object would appear as a flat horizontal line. Deviation from the horizontal indicates dimensions that are changing with time. However, it is important to note that the standard deviation of the CMM is itself approximately 0.0001 in. (≈2.5 μm).

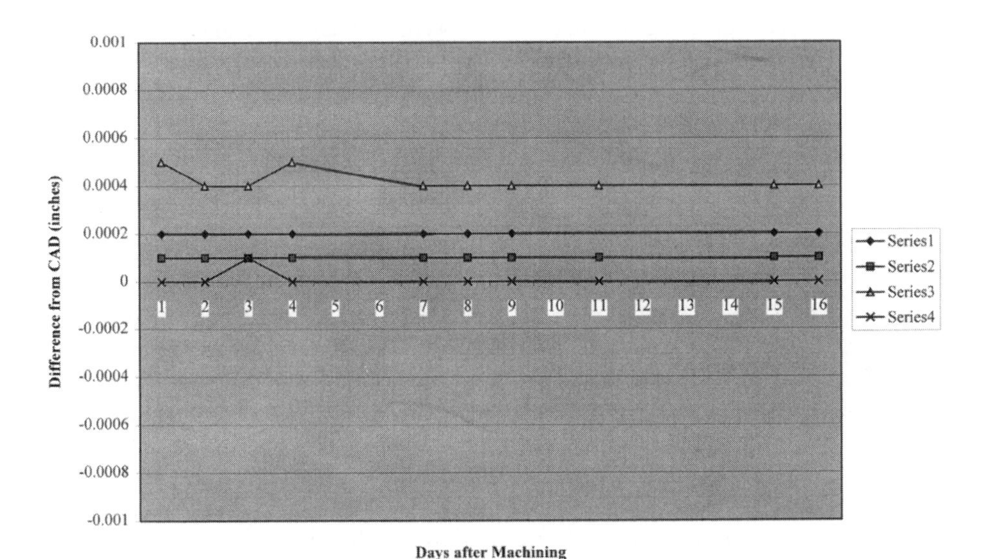


**Figure 4** Dimensional stability data for Ciba Express 2000.

Furthermore, temperature changes in the lab were kept to about ± 1°F, so the maximum thermal expansion/contraction of the material was also of the order of ± 0.0001 in.

Inspection of 14 plots similar to Fig. 4 for all of the 68 measurement traces indicated that *none* of the measurements ever changed by more than ± 0.0001 in. Note that a ''series'' in Fig. 4 refers to the series of measurements at a given location over a period of 16 days. Changes as small as ± 0.0001 in. can therefore be considered to be either (a) within the experimental error of the CMM measurement system or (b) the result of tiny room-temperature variations. Based on these results, we can conclude that Ciba Express 2000 material is exceptionally dimensionally stable under normal laboratory conditions.

The RP&M systems are certainly well-documented means of quickly going from a CAD design to a concept model, especially for complex shapes. However, when exceptional accuracy and surface finish are required, RP&M patterns need to be built with thin layers (e.g., 0.002–0.003 in.) to minimize ''stair-stepping.'' The result is a significantly increased total build time. This occurs because the overhead time (e.g., in SL, the time for the recoater blade to spread a new layer of resin, the time for the system to check the resin level, as well as the ''Z-wait'' fluid relaxation time) is essentially constant per layer. Because additional thinner layers are required for a given part height, the overall build time increases as the layer thickness decreases.

Finally, the time to hand finish an SL part is hardly trivial. Experience indicates that benching SL patterns for use as tooling masters can exceed both the time and the dimensional error budget associated with building the pattern in a SLA (8). One can often CNC a mandrel from an easily machined material [e.g., Ciba REN Shape 540 tooling board, Tool Chemical Co. Die Plank DP-1051, or the recently released Ciba Express 2000 aluminum-reinforced polymer board (9)] in the same overall time required to build and bench an RP&M pattern. Considering the benefits of superior accuracy and dimensional stability, the build times for CNC-machined mandrels are certainly not a liability.

3. The mandrels must now be coated to make them electrically conductive. A number of different methods can be employed. Silver nitrate and reducing solutions can be sprayed with a double-nozzle gun (10). The advantage of this approach is the generation of a very

uniform, extremely thin conducting layer, having a thickness of only a few microns. Clearly, a coating this thin has a negligible effect on mandrel accuracy.

Another method involves simply painting the mandrels with a layer of silver paint. This technique requires special attention to avoid brush marks, and care must also be taken to avoid a buildup of paint in any area. If the paint thickness exceeds 0.001 in., this can affect insert accuracy. Such errors could be important at parting surfaces and shutoffs.

4. The mandrels are now connected as cathodes in an electroforming bath with nickel as the anode. Bath temperature, pH level, current density, chemical concentration, and impurity levels must be carefully monitored and controlled. Electroformed nickel hardness, tensile strength, residual stress, and deposition rate can be varied over a wide range through choice of bath composition and operating conditions.

   The conditions developed by Hasbro/Laser Fare/ExpressTool were established over a period of 6 years through extensive test and evaluation. The resulting ExpressTool electroforming process parameters provide the maximum deposition rate consistent with minimum induced-stress levels. (R. Barlik and T. Feeley, personal communication).
5. After a sufficient thickness of nickel has been electroformed to ensure good abrasion resistance and long tool life (e.g., ~0.080 in. or ~2 mm), the mandrel and its still attached nickel shell are removed from the bath, rinsed in clear water, and subsequently dried. Fortunately, electroforming is a batch process and requires very little labor (primarily to insert mandrels, monitor the condition of the vat, and remove electroformed mandrels when the process is complete). Also, electroforming is not limited by part size. The ExpressTool process has successfully generated inserts up to 30 in. in length, and a tool over 8 ft long is currently in development. The nickel electroforming process generally requires roughly 2 weeks to achieve sufficient nickel-shell thickness. Even 2 weeks is still short relative to the time required to CNC/EDM fabricate conventional core and cavity inserts.
6. The electroformed nickel shell and the attached mandrel are now inverted, and CCCs are positioned behind the nickel surface. The channels are bent to conform to the general shape of the part. Ideally, the CCCs would be positioned based on the results of a thermal

finite-element analysis (T-FEA). The T-FEA would utilize the original 3D solid CAD model to establish that CCC path which results in the *minimum active mold surface temperature difference*, $\Delta T_{max}$. The major benefits of reducing $\Delta T_{max}$ are as follows:

- *Reduced cycle time*. One must wait until the *last* portion of the injected plastic cools below its heat deflection temperature before ejecting the part. Minimizing $\Delta T_{max}$ reduces this wait, decreases cycle time, and increases overall productivity!
- *Reduced part distortion*. Nonuniform active mold surface temperature distributions result in variable part cooling rates and sector-delayed shrinkage, leading to greater part distortion.

In many instances, either schedule or funding limitations make it difficult or impossible to perform a thermal FEA. In these cases, a heuristic approach is commonly followed. Here, either a molder or a mold-maker, or possibly both, will draw upon their experience to establish, in an intuitive manner, the path of the CCC. Having previously encountered problems related to hot spots in earlier tools, presumably they will be able to position the CCC to minimize or at least significantly reduce the magnitude of the most serious hot spots.

7. Next, the combined mandrel, electroformed nickel shell, as well as the positioned and secured CCC are placed in a second electroforming bath. Here, the nickel shell is now electroformed with copper. The reasons for using copper are as follows:

   - Copper can be electroformed more rapidly than nickel, which further reduces tool-generation lead time.
   - As evident in Fig. 1, the thermal conductivity of pure nickel is 88 W/m K. The equivalent value for H-13 tool steel is only 28 W/m K. However, the thermal conductivity of pure copper is 399 W/m K, whereas that for electroformed copper is around 390 W/m K. Although heat conduction in nickel is roughly three times faster than typical tool steels, heat conduction in copper is about 14 *times* faster! Superior heat conduction in the mold leads to faster part cooling, enables earlier part ejection, results in shorter cycle times, and ultimately leads to increased mold productivity!

- Electroformed copper would be too soft for use as the active mold surface. However, it is uniquely outstanding as a *thermal management material*, behind the relatively thin, hard, abrasion- and corrosion-resistant nickel shell.
- The compression strength of electroformed copper is in the 40,000–48,000-psi range.
- Finally, the linear coefficient of thermal expansion for electroformed copper (16.5 μm/m K) is reasonably close to that of electroformed nickel (13.6 μm/m K). The small *differential* expansion (viz., $2.9 \times 10^{-6}$/K) is important since the maximum strains induced during each injection cycle are only of the order of 0.07%, which is well below the yield point of either nickel or copper. Because these deformations are in the elastic region, fatigue effects are greatly diminished.

Significant differential expansion and contraction could lead to substantial induced stress, plastic deformation, fatigue, delamination, and failure of the tool. At this time, Ni–Cu/CCC inserts have already achieved 270,000 shots with no delamination problems. Data for a substantially greater number of injection cycles are currently being developed as a result of an exclusive agreement for joint testing between the General Electric Plastics Division (Pittsfield, MA) and ExpressTool (Warwick, RI).

8. The combined ''mandrel/nickel shell/copper thermal management layer'' ensemble, including the encapsulated conformal cooling channel, is then backed with an insulating material. Once the heat has been successfully transferred to the cooling channels, there is no point in providing high-thermal-conductivity material further into the tool. Thus, a number of commercially available filled epoxy formulations can provide good compression strength with relatively rapid cure (<24 h). When mold pressures are expected to exceed 10,000 psi, machined steel is used.
9. Next, the core and cavity inserts are positioned in a mold frame. The ejector holes are now machined through the backing layer as well as the copper thermal management layer and the nickel shell. Of course, care must be taken to avoid drilling through a conformal cooling channel. In this regard, an additional advantage of T-FEA positioned conformal cooling channels lies in the accurate location of the channels as they appear in the solid CAD model. The CAD model can then be used as an aid in locating ejection pins to ensure

noninterference with the cooling channels. Finally, the conformal cooling channels are connected to the external coolant supply and return lines.

## V. CASE STUDY 1

Figure 5 shows an automotive wire-harness clip. This part was injection molded in nylon using two different molds. First, ExpressTool, Inc. built electroformed Ni–Cu/CCC inserts for United Technologies Research Center (East Hartford, CT). Concurrently, a conventional H-13 steel production mold at United Technology Automotive (UTA) (Dearborn, MI), was also used to produce the same wire-harness clip. The clip is 60 mm long (2.38 in.) by 35 mm wide (1.38 in.) by 30 mm high (1.18 in.). The same mold base was used in both cases.

Two separate cooling channels were dictated by the wire-harness geometry. One CCC was used primarily to cool the central regions of the part, and the second channel cooled the peripheral regions. Although the geometry of these twin channels was complicated, it is important to note that these CCCs are easily fabricated and placed behind the nickel shell prior to copper electroforming. In this way, the cooling channels are completely encapsulated in highly conductive electroformed copper. Thus, the heat from the hot plastic can flow (a) through the nickel shell, (b) through the copper thermal management layer, and (c) directly into the conformal cooling channel, where it is transferred away by convection.

After setup, thermal stabilization of the tool, and optimization of the mold parameters, the measured cycle time for the production H-13/DCC mold at UTA was 21 s. This corresponds to 3600/21 = 171 parts per hour, assuming uninterrupted operation of the injection-molding press. Again, after setup, stabilization of the tool, and optimization the mold parameters, the cycle time

**Figure 5** Automotive wire-harness clip.

for the electroformed Ni–Cu/CCC mold was 12 s, corresponding to 3600/12 = 300 parts per hour, again assuming uninterrupted operation of the injection-molding press. Note that 300/171 = 1.75, or a 75% increase in mold productivity as a consequence of utilizing electroformed nickel–copper core and cavity inserts with encapsulated conformal cooling channels!

## VI. CASE STUDY 2

Figure 6 shows a CAD model of a standard Vaseline jar cap injection molded in high-impact styrene for Chesebrough-Ponds. The performance of an existing H-13 steel mold built with conventional DCC was compared to the performance of an electroformed Ni–Cu tool with encapsulated CCC.

After setup, thermal stabilization of the tool, and optimization of the molding parameters, the measured cycle time for the production H-13/DCC mold was 15 s, corresponding to 3600/15 = 240 parts per hour, assuming uninterrupted operation of the molding press. Again, after setup, thermal stabilization of the tool, and optimizing mold parameters, the cycle time for the electroformed Ni–Cu/CCC mold was 9 s, corresponding to 3600/9 = 400 parts per hour, assuming uninterrupted operation of the molding press. Note that 400/240 = 1.67, or a 67% increase in mold productivity when using electroformed nickel–copper core and cavity inserts with encapsulated conformal cooling channels!

It is clear from these two case studies that the reduction in mold cycle time and the consequent increase in productivity for Ni–Cu/CCC molds relative to conventional H-13/DCC steel molds is dramatic. In Sec. VII, FEA results provide an explanation for these substantial reductions in cycle time, as well as major improvements in mold temperature uniformity.

**Figure 6** Vaseline jar cap.

## VII. FINITE-ELEMENT ANALYSIS

To gain a better understanding of the fundamental phenomena occurring within an injection mold, ExpressTool began working with the FEA/Process Modelling and Optimization group at the National Research Council (NRC) (Boucherville, Quebec, Canada) under the direction of Georges Salloum.

The temperature distributions shown in this chapter were developed through a collaboration between the author and Michel Perrault of NRC. The calculations were based on the latest version of the NRC–FEA code. Starting from a CAD model of a specific part, Perrault developed the geometry of the mold, as well as the geometry of both the DCC and CCC cases. Finally, he used representative thermal and mechanical properties for H-13 steel, as well as those for electroformed nickel and electroformed copper where relevant.

This author believes that if one cannot understand a simple problem, the chance of understanding a more complicated problem is greatly diminished. Thus, the part selected for the initial NRC–FEA thermal analysis is a simple circular disk, 3.00 in. in diameter and 0.100 in. thick. Although the part geometry is flat, it has a round shape typical of molded parts, and also has little intrinsic stiffness, with no supporting ribs or gussets.

Figure 7 is a top view of the two cases evaluated by FEA. The sections are split about a plane of symmetry to save computation time, so one is viewing half of each part. The first case corresponds to an H-13 steel tool with DCC, shown on the left. The second case corresponds to an electroformed Ni–Cu tool with encapsulated CCC, shown on the right. For this case, the

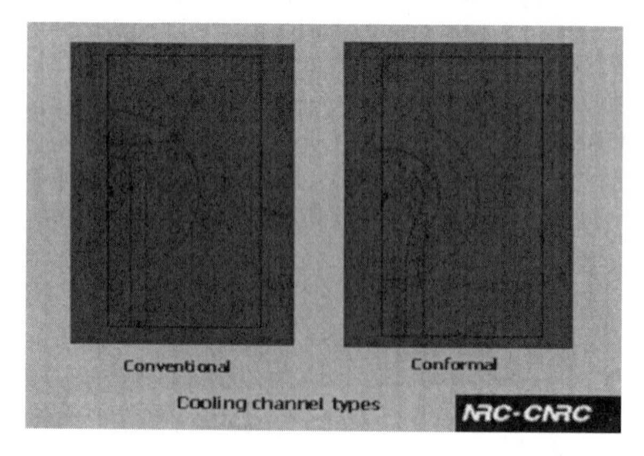


**Figure 7** A conventional H-13/DCC steel mold and a Ni–Cu/CCC mold.

CCC geometry looks something like a "keyhole" when viewed from above. Although, in principle, the CCC could also have arbitrary cross-sectional shape, the channel cross sections were assumed to be circular for this study. In future studies, we will evaluate the effects of noncircular channel cross sections.

Figure 8 shows the model of the Ni–Cu tool developed at NRC by Michel Perrault, which formed the basis of the ensuing FEA analysis. The following assumptions were made:

- The part was center gated.
- The nickel shell was 2 mm (0.080 in.) thick.
- The copper thermal management layer was 4 mm (0.160 in.) thick.
- The copper fully encapsulates the CCC.
- The tool was backed with aluminum-filled epoxy having a thermal conductivity of 2 W/m K.

Note that compared with a thermal conductivity of 88 W/m K for nickel and 390 W/m K for copper, a value of only 2 W/m K for the mold backing material effectively treats the latter as an insulator.

Figure 9 is an FEA image of the distribution of temperature over a cross section through the center of the cooling channels on the core side for the conventional H-13 tool with DCC shown on the left and the Ni–Cu tool with CCC shown on the right.

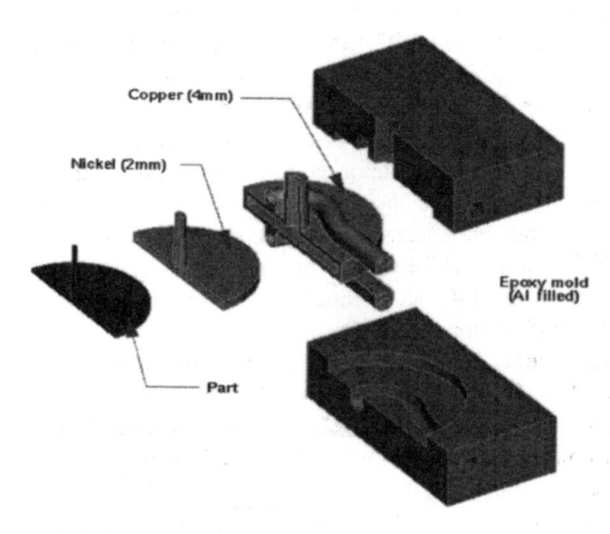


**Figure 8** Model of the Ni–Cu/CCC mold.

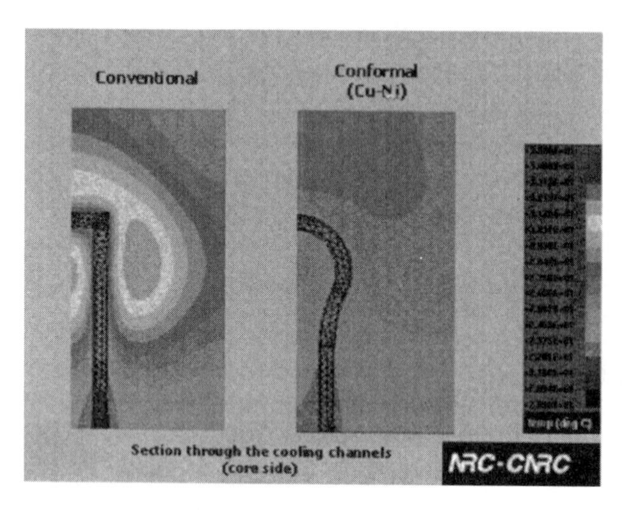


**Figure 9** Core temperature distributions.

The difference in the two temperature distributions is dramatic! The H-13 tool with DCC shows a hot spot to the left of the cooling channel (near the sprue) and another to the right of the channel. Conversely, the Ni–Cu tool with CCC shows an almost isothermal temperature distribution. The value of $\Delta T_{max}$ for the H-13/DCC case is 12.5°C. In contrast, the value of $\Delta T_{max}$ for the nickel–copper tool with CCC is only 2°C. Obviously, the combination of high-thermal-conductivity materials and conformal cooling channels has significantly reduced mold temperature variations in this case.

Figure 10 is another FEA image, this time of the temperature distribution on the *active mold surface* of the cavity side of the tool for the conventional H-13 tool with DCC on the left and the Ni–Cu tool with CCC on the right.

At the active mold surface the effect is even more dramatic. The value of $\Delta T_{max}$ for the H-13/DCC cavity is 18.6°C, and the corresponding value for the Ni–Cu/CCC cavity is only 1.9°C, or, essentially, an order of magnitude reduction in active mold surface temperature variance!

Figure 11 shows the pseudo-color temperature distribution for the cavity surface of the H-13/DCC tool at 2-s intervals from 1 to 15 s after plastic injection. These images illustrate the cooling of the insert over time. Figure 12 shows the same information for the Ni–Cu/CCC tool. It is clearly evident from inspection of these two figures that the cooling rates for the Ni–Cu/CCC tool are much faster than for the H-13/DCC tool. In fact, the temperatures

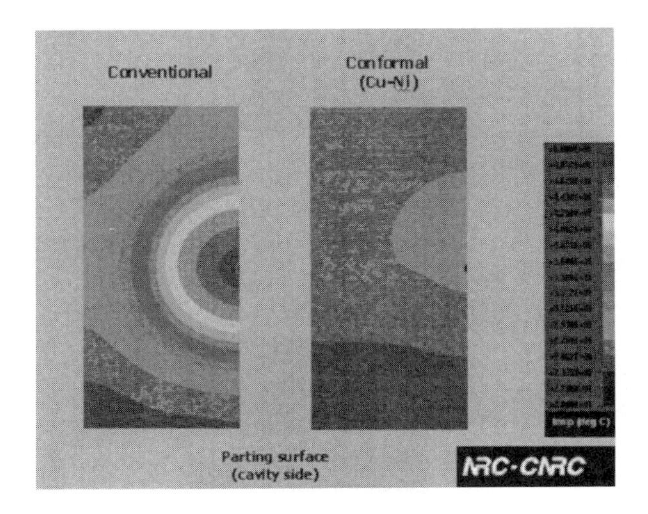


**Figure 10** Cavity temperature distributions.

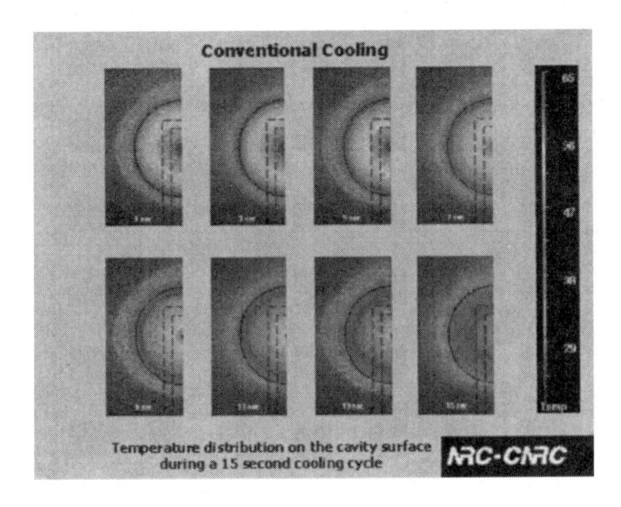


**Figure 11** H-13/DCC temperature versus time.

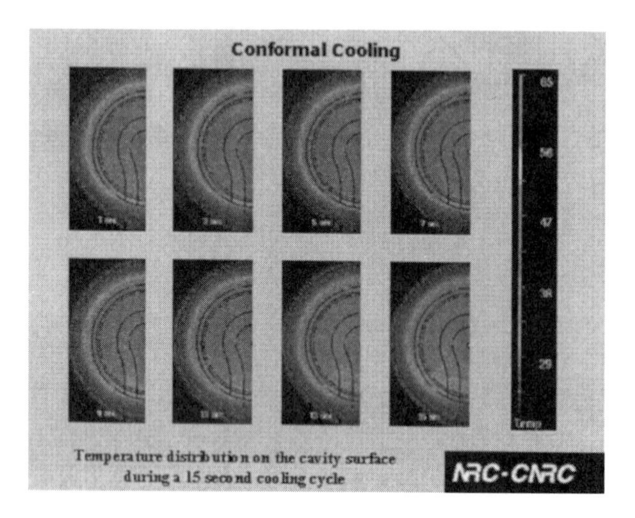


**Figure 12** Ni–Cu/CCC temperature versus time.

throughout the Ni–Cu/CCC tool only 3 s after injection are already *lower* than the corresponding temperatures for the H-13/DCC tool after 15 s!

These data begin to explain the reasons behind the extraordinary productivity improvements noted in the two case studies presented in Sects. V and VI. The only reason the productivity gains are not even greater is that the cycle time includes not only the cooling time but also the times needed to (a) close the press, (b) inject the plastic, (c) pack the plastic, (d) open the mold, and, finally, (e) eject the part. However, neither the thermal conductivity of the mold nor the presence of CCC has any effect on these five time intervals. Thus, the dramatic productivity gains documented for Ni–Cu/CCC inserts are purely the result of significantly reducing the mold cooling time.

## VIII. PROCESS CHARACTERISTICS

Electroformed nickel–copper inserts with encapsulated CCCs provide a number of important benefits, as well as some limitations. These are discussed in this section.

1. *Thermal conductivity*. As noted previously, the thermal-conductivity values of nickel and copper are both dramatically higher than any of the various tool steels. Assuming 2 mm electroformed nickel

and 4 mm electroformed copper, the effective thermal conductivity of the insert is about 180 W/m K, or more than *six times* the thermal conductivity of typical tool steels. Consequently, for the same heat flow, the temperature gradients in the mold need only be one-sixth as great! The result is a more uniform mold temperature distribution and faster cooling.

2. *Conformal cooling*. A key characteristic of the ExpressTool process is the inclusion of encapsulated conformal cooling channels. The primary benefits are as follows:

   - The reduction in the magnitude of ''hot spots''
   - More uniform core and cavity temperature distributions
   - More uniform plastic shrinkage
   - Less stress induced in the plastic part
   - Reduced part strain
   - Reduced part warpage
   - Shorter cycle times

3. *Enhanced productivity*. Actual performance data for a range of part geometries have shown Ni–Cu/CCC productivity enhancements, relative to P20 or H-13 tools, ranging from 20% to 75%. The average improvement in overall mold productivity has been about 33%. Simply stated, enhanced thermal conductivity coupled with the use of encapsulated conformal cooling channels will, on average, enable the production of 133 plastic parts in the same time that a conventional steel tool would generate 100 plastic parts.
4. *Insert accuracy*. Insert accuracy is critical at parting surfaces and at shutoffs. The mandrels are CNC machined, achieving the same accuracy obtained for other CNC-generated objects. Also, electroforming is atomic in nature, regularly replicating mandrel features within 0.1 μm for the production of CD masters. Finally, electroforming involves almost zero mean shrinkage, so the associated random-noise shrinkage errors are virtually nonexistent.
5. *Speed*. Faster spindle speeds, improved cutter path software, and better cutting materials have reduced lead times for CNC-generated steel tooling by 30% over the past 3 years. However, 12–15 weeks delivery is still too slow, as product life cycles shrink. Ni–Cu/CCC inserts for *production* molds require 7–8-week lead times, with 9–10 weeks delivery for a complete tool with ejectors and frame.
6. *Chemical resistance*. The active surface of the tool is electroformed nickel, which is substantially more resistant to chemical attack than

all conventional tool steels. The best of the conventional mold materials used when injection molding reactive plastics [e.g., poly (vinyl chloride)] are stainless steels. Indeed, nickel is used as an alloy ingredient in stainless steel to *improve* chemical resistance. Experience has shown that Ni–Cu/CCC inserts exhibit virtually no signs of chemical attack during the injection-molding process.

7. *Surface quality*. Electroformed nickel surfaces can be highly polished and have been used for many years in the injection molding of plastic eyeglass lenses. Optical quality surface finishes as good as $R_a = 2\ \mu$ in. ($\sim 0.05\ \mu m$) have been routinely achieved on electroformed nickel.
8. *Textured surfaces*. Mold-Tech, Inc. has successfully textured the active electroformed nickel surfaces of Ni–Cu/CCC inserts. According to Mold-Tech, the resultant texturing using their standard procedures was "sharp, well defined, and capable of good depth when needed."
9. *Mold repair*. A truly unique aspect of building production tools through the use of the electroforming process is the capability to "reelectroform." In the event that a glass-filled plastic has gradually eroded any portion of the active surface of the tool, it is possible to simply mask the unworn portions of the insert and then reelectroform the worn surface. Because the nickel electroforming process adds material at about 1 μm every 5 min, it is possible to rebuild worn areas in a very controlled manner. Obviously, if the tool surface is textured, the rebuilt area will also require subsequent texturing. Of course, the same would be true for a conventional steel tool that had undergone weld repair. One major difference, however, is that weld repair involves considerable heat input and the possibility of insert distortion. Conversely, electroforming is performed in a warm bath involving negligible heat loading and essentially zero insert distortion.
10. *Size*. The electroforming process is not fundamentally or intrinsically limited in size by any accuracy, plating, or processing step. Because CNC is certainly capable of producing large mandrels accurately, and electroforming involves essentially zero random-noise shrinkage, the only limit at present involves the size of the vats. The current ExpressTool electroforming vats are about 3 ft wide by 5 ft long by 2 ft deep. This has been more than sufficient for all projects performed to date. Should larger inserts be required,

larger electroforming vats are certainly feasible and could be built, calibrated, and operational within a few months.

There are also a number of limitations to the ExpressTool process. Some of these are cultural and apply to all forms of rapid tooling; others are specific to this process. Among these limitations are the following:

1. *Cultural.* For many molders and mold-makers, the statement ''If it isn't made out of steel, it isn't a production tool'' summarizes their perception. Clearly, this attitude will slow acceptance of *all* new forms of alternative production tooling, in general, and electroformed tooling, in particular. Just as machinists initially resisted CNC, but gradually embraced the new equipment when productivity gains became obvious, this author believes the same shift will happen here. When numerous case studies have clearly and conclusively documented the productivity gains and reduced part distortion, market forces and global competition will pull manufacturing in the direction of lower unit cost and higher part quality.
2. *Tool life.* At present, data regarding the tool life of electroformed inserts for a range of unfilled and filled thermoplastics are incomplete. Although numerous inserts have already run in excess of 100,000 shots and 1 tool has reached 270,000 shots with no visible signs of wear, none of the inserts has yet been run to failure. The reason, simply stated, is that none of the projects responsible for their development have required larger numbers of parts. Recently, G.E. Plastics (Pittsfield, MA) and ExpressTool signed an exclusive joint agreement intended to document the following:
   (1) The *cycle time* for Ni–Cu/CCC inserts versus tool steel inserts
   (2) *Part distortion* with Ni–Cu/CCC inserts versus tool steel inserts
   (3) Ni–Cu/CCC *insert lifetimes*, for glass-filled and neat GE plastics.

   A cycle time as low as 10 s has already been achieved using Ni–Cu/CCC inserts. Assuming 12 h per day, 5 days per week operation of the injection-molding press at GE Plastics, one can mold about 20,000 parts per week, 250,000 parts in 3 months, or 1 million parts in about 1 year. With allowance for finite press downtime, these intervals will probably increase somewhat. As these data are collected and analyzed, it will be made available to the public in the form of publications, mailings, and information posted on the Internet.

3. *Deep recesses*. A fundamental characteristic of the electroforming process involves the transport of ions along electric field lines. When the conducting mandrel is connected to a voltage source, an electric field is established. Because electric fields are stronger near external corners and weaker near internal corners, plating occurs more rapidly near the former and more slowly near the latter. The nonuniformity of the electroformed coating is not a problem itself. However, if inadequate plating time is allocated, then internal corners may not be sufficiently thick to ensure long tool life. Thus, core or cavity geometries involving recesses with aspect ratios (i.e., depth/gap width) greater than 3 require additional nickel electroforming time. When aspect ratios greater than 6 are essential to part function, ExpressTool will generate a machined steel insert rather than attempt to electroplate such a high-aspect-ratio recess. Thus, the electroforming process is ideally suited to smoothly varying, albeit complex, curved geometries and is less suited to geometries involving high-aspect-ratio recesses with sharp corners.

## ACKNOWLEDGMENTS

The author would like to acknowledge the outstanding cooperation of the Process Modelling and Optimization group at the the National Research Council, Boucherville, Quebec, Canada, under the direction of Georges Salloum, and especially the extraordinarily capable and creative efforts of Michel Perrault. Mr. Perrault developed the Finite Element Analysis model for the conventional H-13 core and cavity inserts with drilled cooling channels, as well as the FEA model for the electroformed nickel–copper core and cavity inserts with encapsulated conformal cooling channels. The FEA temperature distributions presented in this chapter were the result of his excellent efforts.

## REFERENCES

1. P Engelmann, E Dawkins, J Shoemaker, M Monfore. Improved product quality and cycle times using copper alloy mold cores. J Inject Mold Technol 1(1): 1997.
2. P Engelmann, E Dawkins, M Monfore. Copper vs. steel cores: Process performance, temperature profiles and warpage. Society of Plastics Engineers, Technical Conference, Toronto, 1997.
3. I Sudit, K Stanton, G Glozer, M Liou. Thermal characteristics of copper-alloy

tooling in plastic molding. Report No. 397, Department of Mechanical Engineering, Ohio State University October 1991.
4. B Bedal, H Nguyen. Advances in part accuracy. In: P Jacobs, tech. ed. Stereolithography and Other RP&M Technologies. Detroit, MI: SME New York: ASME, 1996, pp. 164–180.
5. P Jacobs. Fundamental processes. In: Rapid Prototyping & Manufacturing: Fundamentals of Stereolithography. New York: SME/McGraw-Hill, 1992, pp. 79–110.
6. T Pang. Advances in sterolithography photopolymer systems. In: P Jacobs, tech, ed. Stereolithography and other RP&M Technologies. Detroit, MI: SME/New York: ASME, 1996, pp. 27–79.
7. B Bedal, H Nguyen. In: P Jacobs, tech. ed. Detroit, MI: SME/New York: ASME, 1996, pp. 156–162.
8. T Mueller. A model to predict tolerances in parts molded in pattern based alternative tooling. Proceedings of the 1998 SME Rapid Prototyping & Manufacturing Conference, Dearborn, MI, May 1998, pp. 559–577.
9. K Filipiak. Injection molding thermoplastic parts in days in tooling produced from new composite board. Proc. 1998 SME Rapid Prototyping & Manufacturing Conference, Dearborn, MI, May 1998, pp. 223–243.
10. International Nickel Co. Electroforming with Nickel. American Electroplater's Society, Inc., 1997, pp. 12–14.

# 9

# An Automotive Perspective to Rapid Tooling

**Anthony T. Anderson**
*Ford Motor Company*
*Redford, Michigan*

## I. INTRODUCTION

On a global and domestic scale, America's share of the automotive market has decreased primarily due to increasing foreign competition and rapid market growth in Asia and South America. In 1965, U.S.-based manufacturers produced over 53% of all vehicles sold in the world, with an 8% average return on sales. Today, the United States makes only 36% of all vehicles sold in the world with less than 2.5% return on sales (*Automotive Industries*, November 1997, p. 5). Since the end of the cold war, the U.S. automotive industry has been forced to change to become more competitive in a rapidly growing global economy. The industry has pushed to institutionalize processes that provide speed to the marketplace: simultaneous engineering, agile manufacturing, world-class timing, and corporate globalization. The Japanese have provided the benchmarks for change, where quality and cost competitiveness have become required entry fees to the game. These competitive challenges have put a strain on U.S. automotive manufacturers to maintain their share of the market with a production system that evolved in the absence of these concerns. In response, efforts are being made to incorporate processes that improve communication both internally and within the supplier base, to take full advantage of our diverse workforce and become more flexible as the market continues to become more global. These efforts provide a basis for recognizing potential rapid tooling (RT) applications from an automotive perspective.

Current trends to reduce the product-development cycle time and manufacturing cost in the automotive industry are discussed in terms of (a) our utilization of rapid prototyping and manufacturing (RP&M) to accelerate the product design process and (b) the emergence of rapid tooling (RT) technologies for future low-cost niche market manufacturing. These trends involve the integration of computer-aided fabrication technologies with proven low-cost fabrication processes to develop more economical manufacturing methodologies with improved system robustness. From an automotive manufacturing perspective, successful implementation will rely on our industry's ability to improve communication through cross-functional team efforts while reducing technology development costs through multiple-resource leveraging. Meeting these ''challenges of change'' will be key to survival for the North American Automotive Industry in the 21st century.

## A. Approaching Niche Vehicle Markets

The manufacturing problems associated with future low-volume niche-car market requirements (i.e., making less than 100,000 vehicles) are unique from a U.S. automotive manufacturing point of view. Unlike the rest of the world, U.S. auto manufacturers evolved in a atmosphere where vehicles were produced in high volume. Product development and high fixed tooling costs could be amortized over the production life of many vehicles. New equipment, quick die change strategies, and Just-in-Time (JIT) operations are easily justified in high-volume production (1). Auto sales must be high to offset both the cost of traditional product redesign and subsequent tool fabrication and still make a profit. In contrast, the U.S. aircraft industry developed in an environment where production volumes are relatively low and resulting product costs are relatively high. To be competitive, they resorted to extensive use of computer-aided engineering (CAE) simulation methods early in the design stage of the development cycle to minimize the high cost of redesign. They could not be competitive absorbing both redesign and tool-development cost. They had to ''get it right the first time'' to survive. Today's automotive customers expect more—low volume sales must not imply high product cost. To be competitive in today's niche-car markets, automotive original equipment manufacturers (OEMs) need to meet customer demands by also ''getting it right the first time'' and producing higher-quality products with faster development lead times and at lower tool-fabrication cost. Efforts to address these challenges are reviewed and compared to traditional methods.

Traditional machining incorporates the use of a series of dedicated machines (milling, turning, drilling, boring, and grinding) for material removal.

Even if each machining operation can be done quickly (high-speed machining), setup time and idle time periods between each stage can be extensive. Efforts have been made to reduce these bottlenecks between machining operations by utilizing five-axis machine cells that combines operations (agile manufacturing). These machining systems are more flexible than dedicated machines for part manufacture (2,3). Unfortunately, their associated high variable cost and complex tool path generation make implementation for competitive low-volume manufacturing difficult to justify. The existing problem is that whereas computer-aided design (CAD) can easily design complex parts and today's machine tools can easily and efficiently cut them (high-speed machining), the process by which the multiaxis machining motion is described has not changed significantly for almost 30 years. Despite reports that automated tool path generation has made significant progress reducing product-development lead times (4), Automatically Programmed Tool (APT), the underlying mathematical technology for multiaxis machining, does not meet today's machining needs. Except for specific cases where parts have smooth contoured surfaces (stamping dies), the highly skilled APT programmer must discretely program every surface and check for each potential gouge, tangent, or surface discontinuity. Although APT-based systems can program complex parts, these systems take long times to learn and the programs generated are characteristically complex and are difficult to verify. Nelson Metal Products claims to have developed a time-saving software program capable of generating tool path data directly from CAD data with ''minimal'' human intervention. These tool path data are used to make a complex prototype part by computer numerically controlled (CNC) machining. Nelson uses this RP&M tooling fabrication approach to reduce the lead time for optimizing Ford's Front End Accessory Drive bracket design. Unfortunately, their approach is limited to this specific application. To date, truly automated tool path generation software is being successfully applied only to specific applications. A more general CNC software program has yet to be developed that can generate tool paths directly for five-axis machining of arbitrarily complex parts without some highly specialized, human intervention. In general, NC programming with its associated high level of human interaction remains the major bottleneck in the product-development process.

To date, low-volume product developments have been fueled by a successful systems approach to lowering fabrication cycle time among OEMs in the auto industry. For example, assemblies like Ford's Sheet Molded Composite Aerostar hood are now produced as a single part in a minute or less. Although consolidating this assembly into a single more complex part takes longer, the total fabrication time and cost is far less than what is required to form and join simpler designed components together. Likewise, General Motors' thermoplastic (30% glass-filled polycarbonate/polyester) door module consolidates as many as 61 individual parts (most were metal stampings), reducing assembly time by 84% on their minivans, Chevrolet Malibu, and Oldsmobile Cutlass doors. DaimlerChrysler's Composite Concept Vehicle, once known as the China Concept Vehicle, represents the ultimate in part consolidation. The entire body shell consists of just four injection-molded composite plastic components (15% glass-filled polyethylene terephthalate). These 4 components would replace over 80 stamped and welded parts in a typical steel car body. Also, ferritic stainless exhaust manifolds (currently a stamped, tubular, welded assembly) can be hydroformed with 33% fewer operations and 20% fewer assembly components. In addition, injection-mold and hydroform tooling requirements are much less severe than those for traditionally stamped components, making their use ideal for low-volume fabrication. Other cost-saving fabrication technologies for low-volume component manufacture include reaction injection molding (RIM) and resin transfer molding (RTM). These specialized processes help lower overall cost and vehicle weight for specific part applications such as structural components (underbody cross-members, floor pans, and other body parts) by replacing traditional steel stampings with lighter weight, fiber-reinforced, plastic composites. Although composite parts take longer to make than steel stampings, tool requirements for RIM and RTM parts are much less severe. As a result, urethane or Ni-shell molds with cast aluminum, epoxy, or cement backing can be used for production of composite structural components at a fraction of the cost and lead time of traditionally machined tool steels (5,6). The utilization of shell-mold designs for accelerated tool fabrication is reviewed in Sec. V.

Another approach to being competitive in the niche-car market is to go global. High-speed communication technologies allowed Ford leadership to develop a genuine global car, the CDW27. When Ford of Europe needed a new mid-sized family car, with a market potential of only 25,000 units a year, the company could not make a profit building a sophisticated niche vehicle for one region. However, by spreading the cost of development and production around the world, a 100,000 units per year market potential could be realized, where the high initial cost could be offset. This strategy allowed Ford's CDW27 to become the first modern global car. It was named Mondeo in Europe, Taiwan, and the Middle East, and slightly modified versions went on sale in North America with the names Ford Contour and Mercury Mystique. The success of Ford's CDW27 became the prototype for a new way of thinking about a range of product developments with common platforms. It proved that true globalization was finally possible with enabling communication technologies and that customer-focused teams were the way of the competitive future.

Unfortunately, these efforts to lower overall product-development cost are effective for specific applications only. They still lack the robustness needed to allow us to be more competitive in the niche-car market in general. Some insight to this problem can be made more apparent by reviewing our traditional product-development process and how computer-aided technologies can assist.

## B. Accelerating Product Developments

As seen in Fig. 1, U.S. automotive manufacturers lag behind the Japanese in reducing the product-development cycle. A major proportion of this deficit can be attributed to their reduction in engineering changes (e.g., ''by doing it right the first time''). The Japanese have demonstrated a great willingness to more readily utilize the kinds of technological tools that help reduce cycle time. They have made the most of their common cultural heritage to better communicate and work together. Past experiences have taught them to depend on each other to survive in a global economy. On the other hand, U.S. auto manufacturers and their suppliers developed in an environment in which all

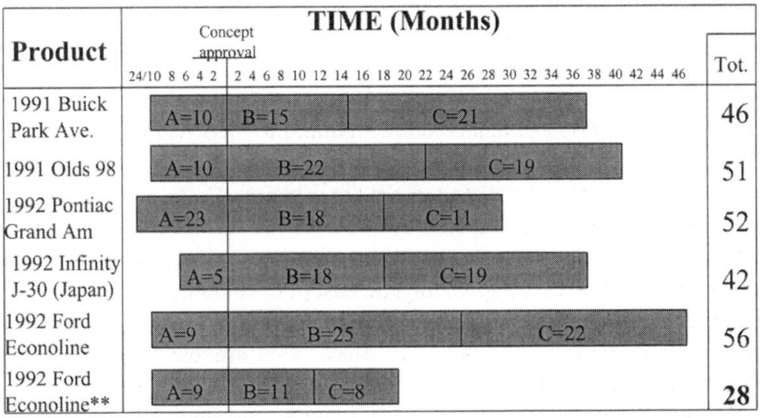

| Product | TIME (Months) — Concept approval; 24/10 8 6 4 2 \| 2 4 6 8 10 12 14 16 18 20 22 24 26 28 30 32 34 36 38 40 42 44 46 | | | Tot. |
|---|---|---|---|---|
| 1991 Buick Park Ave. | A=10 | B=15 | C=21 | 46 |
| 1991 Olds 98 | A=10 | B=22 | C=19 | 51 |
| 1992 Pontiac Grand Am | A=23 | B=18 | C=11 | 52 |
| 1992 Infinity J-30 (Japan) | A=5 | B=18 | C=19 | 42 |
| 1992 Ford Econoline | A=9 | B=25 | C=22 | 56 |
| 1992 Ford Econoline** | A=9 | B=11 | C=8 | **28** |

**Figure 1** Product-development time lines A: concept development; B: prototype development; C: manufacturing development. **Development time minimized to show potential of rapid prototype (RP) utilization. (Data from *Automotive Industries*, September 1991.)

competition was localized within the country. These companies grew by being less cooperative and more competitive. The traditional methods of communication between product and manufacturing engineers became the infamous "toss it over the wall" approach. There was little collaboration in the early stages of the product-development cycle. As the automotive market became more global, customer demands for sophisticated niche cars grew to meet their ever-changing social and environmental expectations while government regulations increased for cleaner air and greater fuel economy. These changes increased vehicle manufacturing and organizational complexities both internally and within the supplier base. Unfortunately, the result was longer product-development lead times and higher product cost.

This situation is changing rapidly among U.S. OEMs in the automotive industry. Conventional thinking, limited to the type of machines and methods used in the past, is giving way to radically new approaches to reducing product-development times. Figure 1 forecasts how the integration of RP&M into the product development process can reduce overall cycle time by over 50%, making a U.S. OEM more competitive than ever. This forecast is based on the accumulated influences of rapid prototyping (RP) on the prototype-development stage and RT on the manufacturing-development stage of the product-development cycle. In general, the walls of communication between product and manufacturing are being broken down in the United States by the use of computer-aided technologies.

Figure 2 is a simplified model representing the industry's major communication stages of product development from concept to customer: (1) concept design, (2) prototype verification, (3) tooling fabrication, (4) manufacturing process feasibility, (5) assembly optimization, and (6) customer approval. Traditionally, product-development communication only flows downstream from concept to the customer. When one stage of the process is completed, information is tossed "over the wall" to the next stage. This "one-way" approach to information flow is characterized by many costly, time-consuming, engineering changes that occur further "downstream," making cost-effective globalization difficult to achieve. In contrast, "upstream" communication flow, like customer-driven concept developments ("listen to the voice of the customer"), helps improve sales, and predicts future markets. Likewise, computer-aided technologies like Design for Assembly (DFA) and Design for Manufacturing (DFM) help improve product quality and reduce manufacturing cost. Additional "upstream" information flow between the manufacturing process and tooling fabrication stages encourages process-driven tool development for reduced fabrication lead time and cost (a rapidly growing future trend).

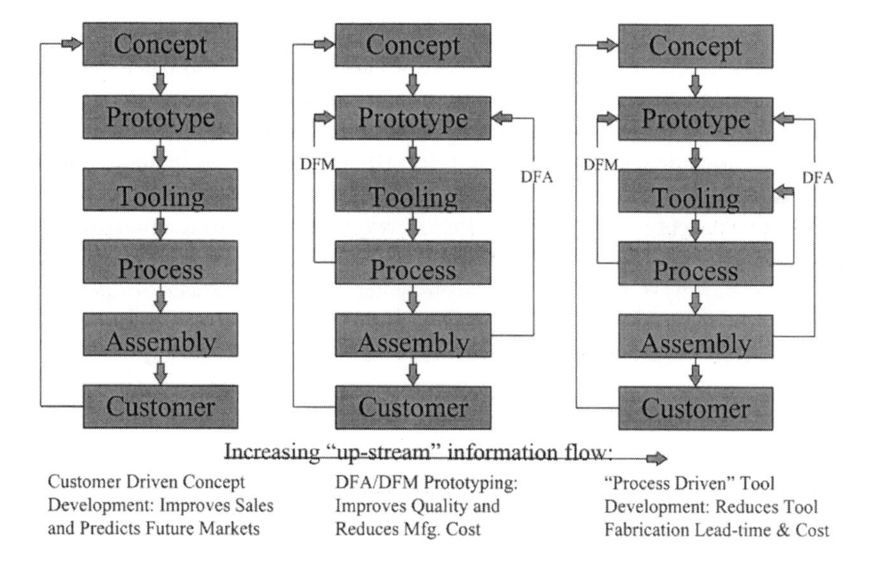


**Figure 2** Concept to customer product-development model (communication to the five development stages improves our competitive edge).

In general, ''upstream'' communication allows engineering knowledge and experiences (things gone wrong) about downstream processes to be made available earlier in the conceptual and prototype design stages. This kind of knowledge-based information flow can help eliminate unwanted engineering changes and rework that would otherwise occur ''downstream'' in the manufacturing-development stages. Traditionally, as much as 80% of total vehicle development cost is built-in during the conceptual design phase. The implementation of computer-aided technologies to improve communication between the six product-development stages would make the overall process more ''seamless'' and flexible for developing a more robust fabrication system. Government-sponsored initiatives like Rapid Response Manufacturing have fueled the emergence of spin-off technologies and development programs designed to address the ''upstream'' communication problem. A 45 million dollar consortium has been formed between GM, Ford, and 10 other high-tech computer hardware and software companies whose primary objective is to develop future computer-aided technologies such as feature-based design, object-oriented methods, and relational database management for accelerating the product and manufacturing-development process.

Reverse Engineering is another computer-aided technology that helps reduce cycle time when redesigns become necessary for improved product quality. Preexisting parts with features for improved performance can be readily incorporated into the desired part design. Reverse Engineering can be used to automatically generate analytical or CAD data representations (point clouds) directly from physical parts, for which no CAD data were previously available. Current applications employ the use of laser scanning (a) to inspect parts where analytical data are generated and can be compared to the original part data and (b) to machine tool inserts where the scanned data are used to generate cutter path data. Unfortunately, part design modifications are often made during sequential processes downstream from CAD operations. Once this happens, the parts made do not match the analytical CAD representation. This problem can be readily overcome with the implementation of "up-stream" communication enablers.

## C. Utilizing Rapid Prototyping and Manufacturing

In one way or another, successful efforts to reduce product-development cycle times in the automotive industry have involved systematic approaches to finding optimal process sequences for utilizing accelerated or rapid fabrication techniques as product-development tools (7–9). Some of the latest developments within the automotive industry have shown how recently emerging rapid prototyping and manufacturing (RP&M) technologies can be used to reduce lead time in the prototype-development process. The term RP relates to a rapidly growing number of automated machines/processes [stereolithography (SL), fused deposition modeling (FDM), selective laser sintering (SLS), laminated object manufacturing (LOM), etc.] which fabricate three-dimensional (3D) solid models directly from CAD image data (either 3D solid or fully surfaced wire frame) automatically, without the use of molds or dies. In general, RP machines utilize two common steps to automatically build a part model: (a) the 3D CAD files are sliced into a series of 2D CAD files that represent cross sections of the part and (b) these "2D" slice files are simple enough to automatically generate the needed vectors to drive the 2½-axis RP machines layer by layer without human intervention. Each layer is built on the preceding layer, by each machine's particular material fabrication technology, until the 3D physical model is built. The process of replacing 3D CAD files with 2D slice files greatly simplifies the tool path generation process over traditional APT machining. In contrast to traditional multiaxis tool path machining, RP machines make part models quickly by avoiding complex tool path generation for a more robust part-building process.

When this effort relates to developing prototype parts rapidly by the strategy of integrating RP technologies with secondary or subsequent operations, this integration process is still commonly referred to as rapid prototyping (RP). Subsequent casting operations have been found to be useful for making prototypes with material properties close to or almost identical with the desired production material. When the objective is to use this part as a tool (mold, die, or fixture) for manufacturing other parts, the process is referred to as rapid tooling (RT). Tool fabrication is a major bottleneck in the manufacturing phase of the vehicle-development process. When product-development cycle time and cost-reduction efforts relate to both RP and RT applications, the process is referred to as rapid prototyping and manufacturing (RP&M). In general, the application of RP&M methods to accelerated product development can be grouped in one of four categories: machining laminates, RP stages, subsequent casting operations and RT developments. In this chapter, these categories are used to describe the influence RP&M techniques can have on the traditional product-development process. Also, a case study overview is presented for an injection-molded part to illustrate the impact RP&M can have on potential lead-time reductions for both prototype verification and tool fabrication.

## II. MACHINING LAMINATES

The distortion problems encountered in state-of-the-art 3D RP patterns have yet to be fully resolved. RP pattern warpage is believed to result from residual stresses created during the laminating build process. These stresses distort the part after postcuring. Even though great progress has been made in this area, further work is still needed before the toolmaking community will fully accept the use of RP models as patterns. Acceptable patterns can be accurately made by machining (i.e., CNC) but, as mentioned earlier, this approach requires costly and time-consuming user interaction. Some progress has been made, but much more is still needed to improve the robustness of the process for making complex parts. Efforts have been made to overcome this shortcoming by utilizing the slice or laminating build feature of RP technology to make automated tool path programming more robust and to accelerate the tool-fabrication process (10–13). In contrast to the RP processes (where materials are deposited in layers), machining laminates involve material removal in layers. Slice or laminated tool path applications to NC machining (of wrought material sheet or plate stock) can be grouped into four categories: precision stratiform machining, computer-aided design/laser-assisted manufacture

(CAD/LAM), profile edge lamination (PEL), and direct slice control machining (SCM).

## A. Precision Stratiform Machining

To make functional prototypes from engineering metal materials, this (computer-aided) slice or layering approach has been successfully applied to traditional three-axis CNC machining. This flexible process—called precision stratiform machining—divides the part into layers (of varying thicknesses depending on the part complexity), generates three-axis tool paths for each layer, and machines each layer to dimension from the desired wrought engineering material stock. This layering approach significantly reduces user interface time for tool path generation, making the process more robust than traditional CNC machining of complex parts. The layers or plates are subsequently stacked horizontally and vacuum brazed together. One of the first components developed using precision stratiform machining was a prototype aluminum cylinder head for Ford's 2-L Zetec engine (14). This part is too complex to be made directly by conventional NC machining from wrought stock. A six-layer working prototype of this complex engine component was made using the precision stratiform machining process in just 100 days, a third of the time required to make traditionally cast cylinder-head prototypes.

## B. CAD/LAM

The CAD/LAM process incorporates the integration of CAD with CAM laser cutting, adhesive bonding, brazing, and mechanical fastening to construct laminated steel molds. This layering approach to traditional machining can also be used to make laminated steel or aluminum tools in a fraction of the time required for traditional machining from wrought or cast metal stock. Ford's Climate Control Division and Toledo Mold and Die have jointly investigated the process for making injection molds. Like the precision stratiform machining process, CAD/LAM utilizes a 3D solid model to define both the part surface outline as well as component details contained within each individual layer. Data for each section are then translated into CNC cutter paths which will be used to cut the outline and internal features of each layer to form stock material. Unlike the stratiform process, CAD/LAM uses sheet stock material of *constant thickness*. Each sheet lamination is thin enough to be easily machined to profile using a laser or traditional CNC machine. The sheets are then stacked laying horizontally, one on top of another, mechanically attached with vertical fasteners, and brazed together by capillary flow of liquid copper between each plate. The steps formed on the mold's surface by the sharp edges of the stacked layers are removed by electrical discharge machining (EDM).

The CAD/LAM process offers a potential material cost and time savings when compared to traditional tools machined from wrought stock. In addition, this laminated-die fabrication approach offers improved cooling capabilities for injection-molded tools. Water lines can be routed to follow, more closely, the contour of the mold's surface geometry and achieve a more uniform cooling of the part (i.e., conformal cooling). Unfortunately, horizontally stacked sheets are difficult to align and secure to one another because clamping must be done through all the sheets to hold them together properly. Also, when thinner sheet stock is used, it has less resistance to warpage during the brazing process. Efforts have been made to replace the brazing process with an adhesive bonding process with some degree of success (15). However, bond strength between layers were found to be directly related to how uniformly the adhesive can be distributed. Using current bonding techniques, it is difficult to maintain an even spread of adhesive with no voids. Further developments are needed in this area before production applications can be addressed for the CAD/LAM process. In addition, the greater the number of sheets used, the greater the problem clamping and brazing the assembly together. This stacked edge bonding problem can be overcome by changing the orientation of the stacked sheets as is done in the PEL process.

## C. Profile Edge Lamination

In a PEL die, the laminations are oriented in the vertical plane and clamped together in a frame. Unlike the horizontally stacked die, the PEL orientation allows the laminates to form a smooth profile across the top of the die face. This leaves the opposite end of the laminates to be used to form a flat face for indexing against a common base plate or vertical wall and to be easily stacked and bolted together to form the completed mold assembly. To date, this approach was applied during a joint collaboration development effort between Simco Industries and the University of Nottingham for making multicavity prototype molds (16). The laminated-mold assemblies were made to form polyurethane foam for automotive door panel insulation at Ford's Utica plant. Laminated molds were made from both aluminum and steel sheet stock (0.10 in thick) by laser cutting the profile of each laminate or part layer. Even though fabrication lead-time reductions would be minimal over traditional machining processes, the laminate feature would readily allow making minor design changes by simply replacing sections of the laminate assembly. In all these cases, however, issues associated with holding the laminated sheets together remain a problem that must be fully resolved before use of these processes become widespread.

## D. Slice Control Machining

This process being developed at Clemson University overcomes the laminate-bonding issue by machining a layer at a time from a solid block of material. This approach utilizes the same robust slice control building algorithms characteristic in most RP machines. The SCM process incorporates the use of a computer hardware/software interface for converting CAD slice files directly into NC code machining data with little to no user interaction. Even though this approach eliminates problems associated with holding the laminated sheets together, the material removal process unfortunately lacks the potential for fabricating improved internal features such as conformed cooling channels in injection molds. Unlike traditional machining, however, the SCM approach simplifies the tool path generation process, giving it the potential of economically machining accurate, quality, prototype patterns and tools faster than current CNC machine technology.

# III. RAPID PROTOTYPE STAGES

In general, the application of machined laminates to accelerate the RP&M process is a step in the right direction, but it lacks robustness. Machined laminate processing limitations (bonding and indexing) restrict its use to specific automotive applications. The demand is increasing for service bureaus that utilize RP technology to incorporate the use of subsequent operations to economically produce multiple copies of functional parts in a more ''productionlike'' material (17). Ford, GM, and DaimlerChrysler have begun to incorporate the use of RP&M as part of their concurrent or simultaneous engineering design cycle. As a result, the number of RP&M service bureaus are rapidly growing throughout the country to meet this demand. In this section, a distinction is made between the various types of prototypes used in the automotive industry. Automotive prototype developments can be grouped into one of two categories depending on the quality of part desired: concept models and functional parts. Depending on desired part volume, these categories are used to describe the impact RP&M technology can have accelerating the prototyping process over traditional methods.

## A. Concept Models

''Touch/feel'' prototypes or concept models are commonly used to communicate design concepts, verify geometric shape intent, and to check some fit issues during assembly with other parts in the early stages of the product-design cycle. Traditionally, model shops work from 2D part drawings to either machine a master part directly or make an ''original model form'' from clay or other soft sculpting material like wood or foam. Design errors are noted and a new or modified concept model is made. This procedure is repeated over and over until a ''visual'' design intent is verified. Accuracy is not a critical requirement for concept models. For small parts, RP models (with tolerances of $\pm$ 0.003 in.) have been successfully used for this application, typically reducing lead time from 8 weeks to 3 weeks. Because RP&M machines work from 3D CAD data instead of 2D drawing data, design misinterpretations are eliminated and undetected human errors are minimized. To date, part size has been limited to the envelop size of the RP&M machine. Larger part models have been made by the assembly of smaller pieces, but time-saving advantages are sometimes offset by the associated design modifications and benching requirements.

Today, model shops commonly use ''cardboard'' composites for rapidly making larger ''touch/feel'' automotive prototypes like interior door panels, instrument panels, and structural body parts. The labor-intensive process has been accelerated by utilizing CAD part data (wire frame or surface) to generate tool paths for driving two-axis NC knife cutters. Automated cutter machines are used to make templates from ''cardboard'' sheets which are assembled to form a 3D ''egg crate'' support and original part model. The part-model template is serrated to conform to the desired part profile when attached to the ''egg crate'' support. The cardboard model is soaked in polyurethane and removed from the support after curing. These models look very ''lifelike'' and can be made in half the time required for traditional clay or cardboard models. When needed, these models are used to make silicone molds for casting more durable polyurethane ''touch/feel'' prototypes.

## B. Functional Parts

''Fit/function'' prototypes are commonly used to verify fit in assembly with other parts and withstand some functional tests in the later stages of the vehicle-design cycle. They are usually made from a material with properties similar to the specified production material and must be dimensionally more accurate than concept models. Traditionally, they are made in small numbers

(10–1000) from ''soft'' prototype molds or dies. With adequate benching, RP&M-built models can be used directly for this purpose for some applications. More often, current RP models are used as prototypes in the early stages of the design cycle. They often lack the desired material properties to be useful for prototype testing. To make functional parts, RP models can be better used indirectly as patterns in conjunction with subsequent operations to rapidly make ''soft'' prototype molds or dies for part fabrication.

In contrast, ''fully functional'' prototypes are commonly used to verify the reliability of the manufacturing process in a production environment (preproduction trials) and occur in the last stage of the vehicle-design cycle. They are usually made in larger numbers (1000–10,000 parts) using ''hard'' prototype molds or dies and are commonly made from the specified production material. In this case, RP&M models can only be used as patterns in conjunction with subsequent operations to rapidly make ''hard'' preproduction tools.

Automotive applications of the RP&M approach for making more durable ''fit/function'' and ''fully functional'' prototypes are outlined in the next section, which describes the most commonly used subsequent or secondary operations for prototype fabrication.

## IV. SUBSEQUENT CASTING OPERATIONS

A number of traditional manufacturing processes integrate well with RP&M technology for accelerating the development of the various kinds of automotive prototypes, including ''touch/feel,'' ''fit/function,'' and ''fully functional.'' Figure 3 is a summary of the most commonly used fabrication methods for making automotive prototypes at various volume ranges (1, 10, 100, 1000, 10000) from a variety of manufacturing materials (zinc, aluminum, cast iron, steel, thermosets, thermoplastics, and elastomers). For metals (aluminum, magnesium, and cast iron), the survey grid shows machining, sand casting, plaster casting, and investment casting to be used most often as subsequent operations for making prototypes. For plastics (injection- and blow-molded thermoplastics and elastomers), machining, vacuum forming, and vacuum and gravity castings were identified as the most important methods for prototype fabrication. In general, the casting processes (vacuum casting, sand casting, plaster casting, investment casting, and spin casting) can be integrated with RP&M technology to rapidly make ''soft'' tools for ''fit/function'' prototypes. These subsequent operations have been found to readily lend themselves to the rapid fabrication of parts from a variety of engineering materials. Their

| Number of Prototypes | Molded Plastics Parts | | | | Metal Alloy Parts | | | |
|---|---|---|---|---|---|---|---|---|
| | Comp. | Blow | Injection | | Die Cast | | Cast | Stamp |
| | SMC | Elasto- | T-set | T-plastic | Zinc | Alum. | Iron | Steel |
| 1-10 | Machine Vac Formed Composites | | Machine Vac Casted Composites | | Machine Sand Cast Plaster Cast | | Machine Sand Cast Lost Foam Hyd. Form | |
| 10-100 | CNC Machine "Soft Tools" | | CNC Machine "Soft Tools" | | CNC Machine Inv. Cast "Soft Tools" | | CNC Machine Inv. Cast "Soft Tools" | |
| 100-1000 | "Soft Tools" | | "Soft Tools" | | "Soft Tools" "Hard Tools" | | "Soft Tools" "Hard Tools" | |
| 1000-10,000 | "Soft Tools" "Hard Tools" | | "Soft Tools" "Hard Tools" | | "Hard Tools" | | "Hard Tools" | |
| 10,000-100,000 | "Hard Tools" | | "Hard Tools" | | "Hard Tools" | | "Hard Tools" | |

**Figure 3** Low-volume fabrication tool grid. ''Soft/bridge tools'' (epoxy, kirsite, Al) and ''hard tools'' (steel, Ni, ceramic) are either machined directly or made indirectly from patterns by either a cast, thermal spray, or deposition operation.

application to RP&M for making plastic and metal automotive components will be briefly summarized.

## A. Rubber Mold Casting

A variety of prototype plastic parts can be developed for injection and compression molding using a vacuum/gravity casting process. RP&M models can be used as patterns to cast silicone [room-temperature valcanizing (RTV)] molds (18) for making polyurethane prototype parts (1 to 30 parts/silicone mold). Vacuum-cast polyurethane can be used to make a variety of plastic ''fit/function'' prototype automotive parts because its hardness can be adjusted to match the corresponding thermoplastic production material. This flexibility also allows polyurethane prototypes to be used in fluid-flow analysis for design evaluations. Vacuum casting polyurethane prototypes in RTV molds can reduce development time by as much as 90% over traditional prototyping by injection molding, where epoxy composite molds (19) or aluminum dies were previously required. In addition, polyurethane core box molds have been made from RP&M patterns in one-fifth the time required to make traditional aluminum core boxes (2 weeks versus 10 weeks). A single polyurethane core box can make over 1000 sand cores.

## B. Plaster/Sand Molding

The die-casting process is one of the most economical ways to make a variety of metal automotive components. Unfortunately, tool steel molds (with associated high fabrication cost and long lead times) are typically used to resist the erosive hot metal flow and thermal fatigue encountered when making die-cast parts. The associated time and money make prototype development for die casting difficult. Thus, for eventual die-cast parts, an alternate prototype-development strategy must be employed to effectively test and validate component designs within required budget and schedule constraints. The conventional prototype processes commonly employed for die casting are gravity casting, machining from die castings with similar shape (when possible), and machining from wrought or sheet stock. Among these three approaches, gravity casting has the greatest potential for ''die-cast'' development using RP&M technology. Unlike machining, gravity casting is economical for low-volume quantities and short lead times. After heat treatment, gravity-cast metal prototypes have properties (like surface finish, yield strength, and ductility) that approximate die-cast parts. RP&M models have been successfully used as patterns for making low-cost sand and plaster molds. These molds require simple core designs and parting planes to remove the patterns from the mold before casting. High-quality plaster and precision sand molds have been fabricated to gravity-cast, thin-walled, ''die-cast'' aluminum prototype parts (control bodies and throttle bodies) at a 60% cost savings over traditionally machined prototypes.

## C. Investment-Cast Molding

For more complex part designs, investment-cast ceramic molds (made by either Flask Casting or through the QuickCast™ process) have been used to as a secondary operation to gravity cast prototype metal parts. The investment-casting process (best suited for fine detail and close dimensional tolerances) can be readily modified to make its use ideal for the rapid prototype development of complex ''die-cast'' parts. The modified process (a derivative of the conventional lost-wax investment-casting process) incorporates the use of cross-linked photopolymer ''QuickCast'' models as expendable patterns. These patterns are burned off during the ceramic-shell mold-making process, instead of being melted away as with traditional wax patterns. No parting lines or cores are needed because the QuickCast patterns are burned out of the mold cavity. As a result, the soft-metal (aluminum) tooling required to make traditional wax patterns is simply not needed. These advantages simplify the prototype-development process for complex parts or where conventional mold designs (with many cores and parting lines) become cost prohibitive. For example, aluminum (SAE 356) prototype reactor blades for automotive torque converters were made within 3 weeks instead of the required 30 weeks for conventionally machined prototype reactor blades. The parts satisfied the specified tolerance of 0.02 in.

Furthermore, a tolerance of ± 0.002 in. (approximately 0.05 mm) have been obtained for small parts by investment casting. Parts can range in size from 1 to 36 in. The larger the part, the greater the tolerance required by the process. In general, smoothness and accuracy becomes increasingly difficult to obtain as the size of the casting increases.

## D. Spin Casting

Spin casting can be used to make plastic, wax, or soft-metal prototypes in sizes smaller than 9 in. The spin-cast process consists of pouring molten metal or liquid thermoset plastic resin into the center of a spinning (200–1000 rpm) vulcanized silicone mold. This rotation forces the material outward under centrifugal force, resulting in pressures of 10–15 psi which distributes the material throughout the mold cavity and expels any trapped air before solidification occurs. Metal parts can be spin cast at 50 casting cycles/h. Plastic (thermoset) parts can be cast at 10–15 cycles/h. The number of parts that can be made per cycle can range from 1 to 10 parts, depending on part size and mold size. Surface finishes of 90 μm root mean squared (rms) are possible, and casting tolerances of 0.005–0.008 in./in. can be maintained from part to part. The two preferred silicone materials used for tooling in spin casting are room-temperature-cured RTV and heat-cured vulcanized rubber. The RTV molds can withstand temperatures as high as 600°F, whereas vulcanized rubber molds can withstand temperatures as high as 1000°F. Unfortunately, vulcanized rubber molds are formed under pressures as high as 4000 psi and at temperatures as high as 400°F. Thus, RP&M parts to be used as patterns must be able to withstand this temperature. Common practice is to use pewter or high-temperature plastic parts produced in RTV molds as patterns or submasters to create multicavity heat-cured vulcanized rubber molds. Any metal that melts below 900°F can be readily spin cast in vulcanized silicone molds.

## V. RAPID TOOLING DEVELOPMENTS

Making tools for both prototype part development and production component manufacture represent one of the longest and most costly phases in the automotive product-development process. The sequential approach to production tool fabrication by conventional machining is characterized by long lead times and high cost. As a result, current practice is to start tool fabrication long before product design is complete. Unfortunately, late design and engineering tool changes commonly occur, making tool-fabrication lead time unpredictable. For example, front and tail light reflector molds may undergo as many as 16 tool-design alterations before completion. To accommodate these changes, the tool material must not only be relatively soft to readily remove material but also must be weldable to add material when needed for design changes. Conversely, the tool material must be hard enough to resist wear and forming loads. Traditional tools (molds, dies, and fixtures) are machined from wrought tool steel billets. Tool steels like SAE 4340, H-13, and P-20 are most often used as die materials in production because of the unique properties obtained through alloying and heat treatment. As a result, traditional tool-material selection is usually a compromise of properties (machinability and weldability versus wear and strength) affecting performance.

This machinability compromise can be minimized using RT&M fabrication methods. RT&M can have a significant influence on reducing product-development cycle time and cost. Figure 4 shows potential lead-time reductions of various RT technology categories for low-, medium-, and high-production-volume applications. Time-reduction estimates were based on the

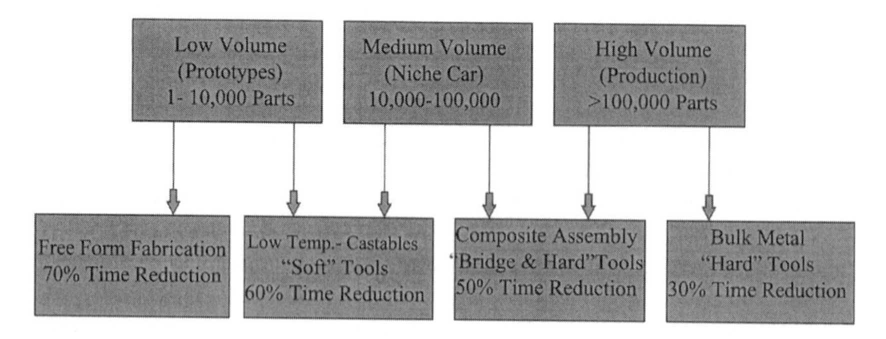


**Figure 4** Rapid tool development. Rapid tool fabrication methods are categorized for various part volumes and reflect potential lead-time reductions by eliminating the need for many traditional machining operations.

influence RT&M methodology has on accelerating various tool-fabrication types. RP&M and low-temperature castable tools can be rapidly made for many prototype developments, especially for injection-molding and stamping applications. For niche-vehicle volumes, castable ''soft'' tools, ''bridge'' tools, and composite assembly tools can be used. For higher production volumes, either composite assembly ''hard'' tools, ''bulk-metal'' deposited, or formed ''hard'' tools are applicable. Processes like SCM, EDM, metal spray, and laminated-tool assemblies are grouped into the bulk-metal tool category.

## A. Direct/Indirect RP&M

Stereolithograpy (SL) molds have been used directly to injection mold thermoplastic prototypes in small quantities (20,21). RP&M technologies like 3D™ printing, LOM, and SLS has been successfully modified to make RT directly from metal or ceramic powders (22–25). The 3D printing process involves the selective coating of powder metal (PM) with a organic binder by ink-jet spray. This process is repeated layer by layer, forming a 30-μm particle-sized PM preform directly from a CAD model without molds or forming dies. Recently, the process had been applied to making small injection molds with conformal cooling lines for improved thermal management (26). For LOM, the RT process involves the use of a modified Helisys 1015 or 2030 machine, where a tape-cast powder metal or ceramic sheet is cut by the machine's laser beam to create each cross section. The cross sections are then stacked together. The scrap material is removed and the laminated part is presintered to burn off the wax binder. Final densification is obtained in a conventional sintering furnace. The process has been used to make small alumina wear inserts for composite injection molds and has shown potential for other RT applications.

For SLS, the RT process uses DTM's Sinterstation 2500 System, where metal powders coated with a wax binder are joined together when heated with the machine's modulated laser beam. The resulting porous metal preforms are infiltrated with molten Cu in a batch furnace, creating a fully densified composite metal part with moderate shrinkage. PM steel powders have been used in the SLS process to rapidly make small injection-mold prototype tools that replicate the actual manufacturing process using production materials.

Also, RP&M technologies have been used indirectly as patterns for making molds. For example, low-temperature castable molds like epoxy, chemically bonded ceramics, and RTV silicone rubber can be readily used for making prototype parts (in low volumes). RT methods in this category are made most effectively when RP&M models are used as patterns (27). For intermediate and high production volumes, RT methods can be effectively used to make

PM molds. PM preforms can be made by compacting a mixture of metal or ceramic powders in an intermediate RTV mold that was cast to shape against an RP&M pattern. PM molds have been successfully made from these preforms by three densification methods: Cu infiltration (Keltool™), PM casting, and PM forging. For the Keltool process (28), the resulting ''green'' metal preforms are removed from the mold and sintered to densify the mold part to 70% theoretical. The sintered PM mold is then infiltrated with molten Cu to fill the 30% void space. For PM casting, metal powder preforms are compacted to high density by vibration for reduced shrinkage (less than 0.1%) and improved dimensional control during sintering (29). For PM-forged molds (30), ceramic punches are formed in the RTV molds used to densify PM tool steels by hot pressing. PM molds have the advantage of better heat transfer and die life over epoxy molds. For injection molding, this advantage allows making prototypes in tools that more closely behave like production. For larger parts (greater than 1 $ft^3$), dimensional control becomes very difficult to maintain because of inherent density gradients in the preform and furnace-temperature variations that occur during the sintering or hot-pressing operation. To maintain accuracy for larger parts, machine stock must be added to allow for resulting distortion errors and increasing process time and cost.

## B. Composite Mold Assemblies

For intermediate and high production volumes, making a protective metal shell with cast aluminum, epoxy, or cement backing has shown significant promise for reducing lead time by over 30% compared to traditional tool-fabrication methods. This approach allows mold components to be quickly assembled in a composite structure for improved performance, as shown in Fig. 5. During assembly, the protective metal shell can be mounted to a prefabricated modular steel frame with standard insert bushings and guides for ejection pins and cooling lines. The mold's components are held together by casting, in place, a composite aluminum-filled epoxy (CAFÉ) or cement backing material which supports compressive loads and transmits tensile loads to the frame. Unlike traditional tools, these mold components can be fabricated concurrently and assembled quickly to produce fully functional tooling.

For wear resistance, the mold's surface can be readily cast from any metal material using wax or QuickCast RP parts as patterns. For small tools, shell patterns can be quickly made by any of the RP&M technologies for investment casting the mold's active wear surface to shape with minimal machining (31,32). However, dimensional limitations should be considered when investment casting large complex steel shells. The investment-cast ceramic

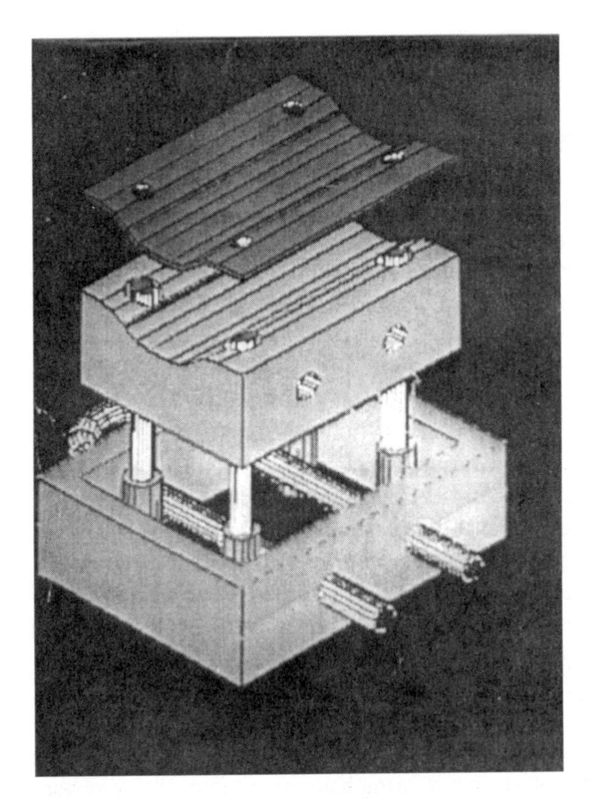

**Figure 5** Composite tool assembly features/components: (from top to bottom) wear-resistant surface shell or coating (arc metal spray, NVD, etc.); castable backing or support material (chemically bonded ceramic, epoxy, etc.); prefabricated inserts (ejection pins, cooling coils, etc.); prefabricated supporting frame (steel, aluminum, etc.).

mold is a thermal insulator and must withstand large thermal gradients generated without failure as the casting cools. These gradients increase in severity as the metal thickness increases and as the ceramic shell's cross section decreases. More heat must be dissipated in heavier metal sections to maintain cooling rates similar to thinner sections. In addition, thinner cross sections in the ceramic shell are less able to absorb and dissipate this heat. This uneven cooling becomes more severe in larger castings, which result in nonuniform shrinkage, dimensional errors, undesirable warpage, and distortion.

As an alternative to investment casting for making larger molds, RP&M patterns can be used as mandrels for depositing wear-resistant Ni on a molecular level to form high-density-shell tools with great dimensional accuracy, ideal surface finish, and minimal distortion. This can be done by nickel vapor deposition or electroforming. Other applicable shell-making technologies that can incorporate the use of RP for RT composite mold assembly developments are arc metal spray, bulk metal spray and EDM. Because machining is minimal, tool life can be improved by selecting protective shell materials with greater wear resistance and heat-checking resistance than conventionally used tool steels. The most significant of these shell-fabrication methods (for making large tools) are described in relation to their applications to niche market manufacturing.

## C. Arc/Bulk Metal Spray Shells

The arc metal spray process can be used to deposit low-melting alloys (below 800°F) directly onto plastic RP&M patterns, forming a solid metal impression of the pattern without machining. The process uses two spools of metal wire as consumable electrodes which are fed through a spray gun. An electric voltage is applied across the electrodes, forming an arc which melts its tips. Molten droplets are forced off the electrodes and cooled by a high-velocity airstream which propels them toward the pattern substrate, forming a mechanically bonded coating subsequent to impact and cooling. When properly sprayed, the shell coating formed never reaches temperatures above 120°F, making RP&M patterns ideal for use as substrates. This process continues, layer upon layer, forming a solid metal shell conforming to the profile of the pattern shape. Arc metal spray tooling is generally produced by spraying a soft-metal shell (kirksite, a zinc-based alloy) an inch thick or less and backing up with CAFÉ or a ceramic (33). The pattern is then removed from the metal shell after the mold is constructed. Unfortunately, the protective metal coating produced in this way is porous and is susceptible to flaking and spalling during service. As a result, spray metal molds are best suited for applications requiring low pressures and temperatures (vacuum forming, blow molding, RIM, injection-molding polystyrene patterns, sand core boxes). Arc metal spray molds can be made in a fraction of the time required for traditional cast kirksite prototype molds when using RP&M models as patterns.

Unlike porous arc metal spray coating deposits, bulk metal spray processes are characterized by high deposition rates where porosity and residual stresses are minimized (34). For example, ''simultaneous spray peening'' and the Osprey process can atomize as much as 150 lbs/h and 200 lbs/min, respectively, of molten metal. In the former process, metal deposits are simultaneously shot peened to increase density and reduce residual stresses. Shot peening can be combined with a variety of thermal spray techniques (induction melts, wire arcs, etc.) to control warpage resulting from internal stresses that are created from thermal gradients inherent to the process. The major drawback is the large quantities of shot required which limits throughput when high deposition rates are desired. In contrast, the Osprey process is commercially used to make large billets, tubes, and sheet stock weighing several tons for a variety of specialty steels. The Osprey process produces metal deposits from a induction melt that is atomized and cooled by a high-velocity gas stream which propels the semimolten metal toward a pattern substrate. The process's high deposition rate increases substrate temperatures to 1800°F or more, causing bulk densification on impact. The resulting deposits have low residual stress and a fine-grained microstructure. To withstand the high deposition temperatures, castable ceramic materials have been used as substrate patterns where bulk metal deposits accurately conform to the ceramic substrate's profile. Ceramic substrate patterns can be readily cast from RP&M models with reasonably good dimensional accuracy.

The bulk metal spray processes have been used to make metal stamping tools from SAE 1080 and A2 steel. The Osprey's high deposition rate make its use ideal for large RT fabrication. Unfortunately, its resulting high substrate temperatures produce unacceptable surface porosity at the ceramic/metal deposit interface when spraying on fused silica. Fused-silica substrates can provide fine surface finishes, requires no firing, and have good dimensional stability at elevated temperatures. This problem is greatly reduced for the peening process because its inherently lower deposition rates result in lower substrate temperatures (1000°F). In general, stamping tools made by the bulk processes have been found to (a) reduce fabrication lead time by 80% (1 week versus 5 weeks) over traditional CNC machining and (b) withstand the high-impact loads for good tool life. Over 16,000 stamped parts were made without die failure.

## D. Nickel-Shell Vapor Deposition/Electroforming

Nickel vapor deposition (NVD) is a chemical vapor process that involves the deposition of high-purity nickel directly from a gas vapor. The chemistry of the NVD process was originally developed in the 19th century to purify nickel and make refined nickel pellets and powders on a large industrial scale. Carbon monoxide (CO) gas is passed over nickel powder to form the metastable nickel carbonyl gas, $Ni(CO)_4$. When heated, this toxic metastable gas readily decomposes into its original components (CO gas and solid Ni on a molecular scale). The process has been used to make nickel shells with low residual stress by having nickel carbonyl gas flow over a heated mandrel. Heating the mandrel substrate to temperatures between 110°C and 190°C results in a uniform layer of nickel being deposited on the mandrel at rates between 0.002 and 0.030 in./h, respectively. The resulting Ni shell conforms to the shape of the mandrel, with excellent surface replication, uniform wall thickness, and low residual stresses. Uniform wall thickness is obtained by minimizing mandrel temperature variations. Mandrel materials with high thermal conductivity like aluminum or copper are best for this application. Using the NVD process, a complex Ni-shell mold 1.0 in. thick can be made within 34 h.

This process has been applied to making composite tooling at a fraction of the time for making traditional cast and machined prototype kirksite molds (35). For example, a traditional kirksite injection mold for a plastic automotive instrument panel would cost about $1,000,000 and require over 30 weeks to build. In contrast, NVD or electroformed composite mold assembly for the same instrument panel would cost about $300,000 and require 14 weeks to fabricate. In addition, a traditional kirksite mold can make no more than 50 injection-molded instrument panels before reworking, whereas the harder Ni-composite mold assembly can make over 10,000 parts before rework. This higher-tool-life mold is as good as the more expensive P-20 steel mold traditionally used in production. The high tensile strength (198,700 psi), hardness ($R_c$ 48), and melting temperature (2647°F) of NVD shells make its use as a mold applicable to many fabrication processes: compression molding of SMC, high-pressure injection molding of thermoplastics, RIM of thermoset plastics, and sheet steel hydroforming.

An NVD composite assembly injection mold was made for Ford's SN-95 (Mustang) instrument cluster lens. A traditionally machined P-20 steel to produce an automotive lens tool cost about $120,000 and take 18–22 weeks to make. The NVD tool cost 30% less and had a lead-time reduction of 60%. A 5/8 in.-thick NVD shell was formed over a mandrel CNC machined to the desired shape. After the Ni shell was removed from the mandrel, Cu cooling lines were mounted to its back and added to a premachined platen assembly complete with ejection pins. Cost and timing could be further reduced by replacing conventionally machined A1 mandrels with RP&M mandrels with similar thermal conductivity. The mandrel material must be thermally conductive and withstand deposition temperatures of 350°F. Proprietary epoxy/graphite composite materials have been developed to meet these requirements but have yet to be made directly by a RP&M process. Currently, epoxy/graphite composite mandrels are cast to shape at room temperature in silicone molds (made from traditional RP&M patterns).

Computer-aided engineering simulation indicated temperature variations as small as 2°F across the surface of the NVD tool. Traditionally machined steel molds vary 10°F to 15°F across the surface. This lower temperature variation minimizes part distortion caused by residual stresses, thus improving overall part quality. Over 19,000 parts were successfully made on a 500-ton injection-molding press. No die wear was noted. Mold temperatures were held at 140°F for a 48-s semiautomatic cycle time to simulate current production cycles using conventional tools. The parts showed no warpage as predicted by the CAE analysis. Lowering mold temperature to 120°F during the cycle reduced the cycle time to 30 s (a 30% reduction in cycle time), which greatly lowers piece price. The low thermal mass of the Ni allows the tool to be thermally cycled between 140°F and 120°F for the best conditions to make a high-quality automotive lens. This kind of rapid thermal cycling is not possible using traditional steel molds.

Unfortunately, the NVD process is not readily available because of the potential health hazards associated with using the very toxic metastable nickel carbonyl gas. Even though Canadian companies like NTT and Mirotech have made great progress developing and implementing safety features for the process, more work is needed before use of the NVD process becomes widespread for mold fabrication.

Electroformed Ni-shell tools have been made to injection mold plastic parts (36,37). Electroformed Ni shells have been produced by both ExpressTool, Inc. and CEMCOM Corporation to replace NVD shells as the wear-resistant face to their composite mold assemblies. Electroforming is commonly used in industry as a metal-plating process and is readily available for making Ni shells. Electroforming Ni shells involves appling an electrical voltage between a Ni anode and a cathode (with the desired tool shape), suspended in a aqueous Ni salt bath. The positive Ni cations in a plating bath are attracted to the negatively charged cathode, plating the substrate cathode with a Ni-shell coating. The resulting high-density Ni shell conforms to the shape of the mandrel. Mandrel materials for electroplating must be conductive, insoluble in the plating bath, and withstand deposition temperatures of 130°F. Electrodes have been sucessfully made for electroplating by coating nonconductive RP&M models with a very thin conductive layer of silver or graphite (5 μms).

Unfortunately, deposition rates are low and wall-thickness variations are great, making process implementation slow for composite mold fabrication. Ford Research Laboratories is addressing these issues and a greater use of electroforming for mold fabrication is likely in the near future. ExpressTool has recently announced commercialization of an electroform-based composite tool that utilizes conformal cooling. Reductions in injection-molding cycle times of 20–45% have been demonstrated (37).

## E. Electrical Discharge Machining

Electrical discharge machining is currently used to make complex forging dies. Developed in the 1950s, the process is most capable of handling difficult to machine metals or features like irregular-shaped holes. Unfortunately, many graphite electrodes are normally consumed to maintain accuracy for most mold-making EDM applications. Consequently, traditionally CNC-machined graphite electrodes are used to burn the tool only in the final stage of the machining operation. Also, the fabrication of complex electrodes requires long lead times and high assembly costs. In the early 1960s, a faster, more economical abrading process was developed for making graphite electrodes. The abrading process involves the use of a hard SiC grinding stone (or abrading die) to cut an accurate reverse image of the stone into a block of graphite. The abrading die is attached to the ram of a press and forced onto a graphite block mounted on a table that vibrates 0.020 in. in a orbital motion at over 800 rpm. A graphite electrode of almost any complex shape or size can be produced by this process. Using abraded full-cavity electrodes, the EDM process allows for improved die repeatability, tolerances, and surface finishes.

Rapid prototyping and manufacturing models of the part can be used to replace traditionally CNC-machined patterns to make the abrading die economically. A mixture of silicon carbide powder (120–340 grit) and epoxy resin binder is cast over an RP&M model, which includes the parting line of the electrode, to produce a molded abrasive die-cutting master. These abrading dies can be made in only a day from a finished RP&M pattern. Care must be taken when removing the pattern from the abrading die to avoid damage. The accuracy of the electrode is limited to the accuracy of the original pattern. Fine detail resolution is obviously limited by the orbital motion of the abrading die.

# VI. CASE STUDY OVERVIEW

A case study was made on a small (6 × 5 × 1.5-in. envelope), injection-molded, polyproplene plastic automotive part. This part was an interior cover for the electric sideview mirror—left and right—for the 95/96/97 Ford Contour and Mercury Mystique (see Fig. 6). As a benchmark, lead times were

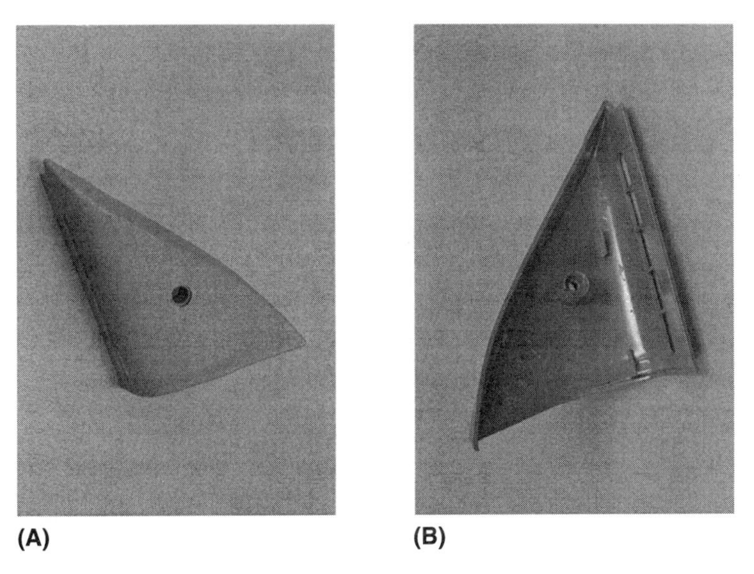


**Figure 6** Injection-molded acrylonitrile–butadiene–styrene interior cover for the electric sideview mirrors on 95/96/97 Ford Contour and Mercury Mystique. (A) Front view; (B) back view.

determined for developing the parts by both traditional and rapid-prototyping methodologies.

## A. Prototype-Development Methodology

Figure 7 is a schematic diagram representing the traditional prototyping methodology for automotive product development. After the concept-design stage, 2D drawings are sent to model or pattern shops to make 3D ''touch/feel'' prototypes. These models are used to verify design intent and as patterns for ''soft''-tool fabrication. If a design error is noted, the 2D drawing is modified to show the design change and another ''touch/ feel'' prototype is made. This design–change iteration continues until the design intent is verified. Once verified, ''fit/function'' prototypes are made using ''soft'' prototype tools. ''Soft'' tools are usually machined from near-net-shaped castings of low-melting alloys like kirksite or aluminum. Prototypes made from ''soft'' tools are used for test to verify engineering design specifications. If the prototypes fail the test, a finite-element analysis is sometimes considered before design changes are made and the whole process is repeated again. This phase of the process

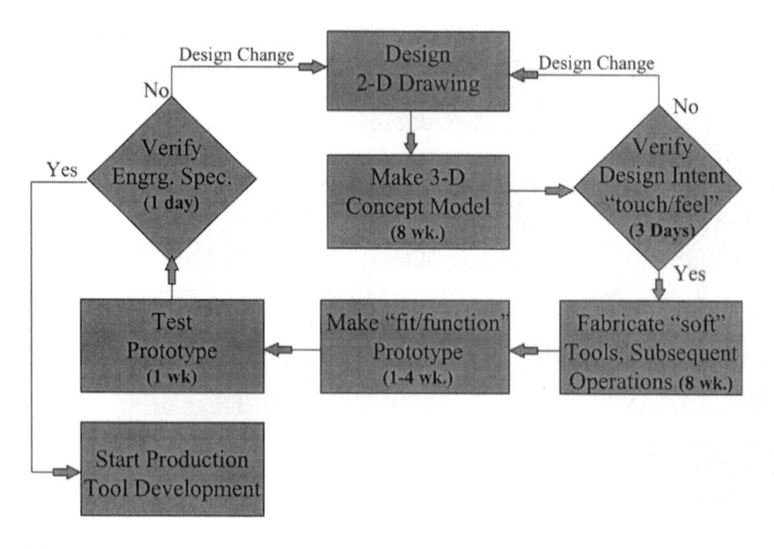


**Figure 7** Traditional prototyping methodology. Nominal lead time: 30 weeks, assuming one redesign.

is characterized by costly, idle time between process steps. When test results meet specifications, the 2D part drawings are ''tossed over another communication wall'' to a tool and die shop where ''hard'' tools are designed and made. ''Hard'' tools are usually machined from wrought tool steel stock like H-13, 4320, or P-20. These tools are used to make either ''fully functional'' prototypes for preproduction trials or actual parts for production.

In contrast, Fig. 8 is a schematic diagram representing the RP&M methodology for accelerated prototyping. The attached diagrams show that using RP 3D models both as patterns and as ''touch/feel'' design correction helps reduce lead time from 30 weeks to 13 weeks (a 55% reduction) for developing the injection-molded prototype design of the polyproplene ''sail'' part. In other words, replacing traditionally made patterns with RP 3D models not only reduces the lead time to make patterns but also allows for comprehensive design evaluations (a time-consuming iterative process) to be made early in the development stage for improved design quality in a fraction of the time normally required for traditional prototyping. Once this iterative design process has been completed, ''fit/function'' prototypes can be fabricated for product verification testing. With the conventional approach, the only opportunity to make design corrections is further downstream after the prototype parts are fabricated.

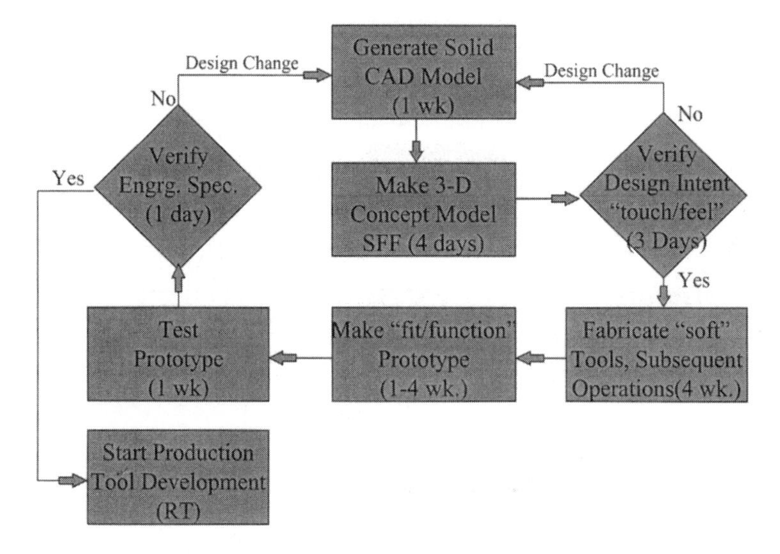


**Figure 8** Rapid-prototyping methodology. Nominal lead time: 13 weeks, assuming one redesign.

In addition, CAE integration into RP methodology occurs naturally because of the readily available computer software programs for automatic grid mesh generation using solid CAD data. RP methodology incorporates the creation of a solid CAD model before 3D part models are made. Thus, the development of finite-element analysis (FEA) can occur simultaneously with 3D model fabrication. Prototype tests results can be compared to FEA results to verify the analysis. Once verified, the model can be evaluated by an optimization computer program to select the best part design instead of relying on the conventional trial-and-error approach normally used. Future improvements in CAD visualization packages and virtual-reality software will allow designs to be more readily understood without the need for physical models. Also, growing confidence in FEA stress analysis will replace the need for mechanical and flow evaluation tests on physical components. Unfortunately, automatic mesh generation is not always best. Depending on the features of the part design, intervention by a skilled FEA modeler is often still required. Eventually, automatic mesh generation will more readily account for various design features, making user intervention unnecessary. Until then, RP will continue to be useful for accelerating the process of making functional prototype parts for test evaluation studies.

## B. Tool-Selection Table

Figure 9 shows tool-fabrication cost and time for various tool/material types. In this table, the influence of RP&M on traditional tool selection is summarized for injection molding the small polyproplene ''sail'' part described earlier. Based on projected tool life, fabrication cost, and lead time, the table optimizes the number of parts to be made for various tool/material types. This optimization table can only be applied to the specific part selected. Even though the table shows a general trend for injection molding most plastic parts, caution should be taken when using this table to extrapolate optiminal tool selection for other parts with different design features, sizes, or materials.

Using the traditional approach, machining is the most economical way to make less than 500 parts. Machined plastic prototype parts are not completely representive of injection-molded parts. They can only be used for simple mechanical test evaluations. In practice, more elaborate prototyping evaluations are delayed until injection molds are made further downstream in the product-development process. As the desired part volume is increased to 3000, the use of cast soft tools (kirksite or aluminum) for injection molding becomes the optimal choice for plastic part fabrication. Cast soft tools cost less to fabricate than hard production tools ($30,000 versus $60,000) even though lead times are almost the same (12–14 weeks versus 16–18 weeks, respectively). For

| Desired Part Vol. (Range) | Optimal TOOL Selection (Traditional Vs Rapid) | | | |
|---|---|---|---|---|
| | Type / Material | Fabrication | | Life* |
| | | Cost ($) | Time (WK.) | Parts/ Mold |
| 1-500 | Parts (machined) | 1000-25000 | 6-8 | ----- |
| 500-3000 | Kirksite(cast) | 25000 | 12-14 | 1500 |
| 1000-3000 | Al (Cast) | 30000 | 12-14 | 2000 |
| 3000-250K | Steel (machined) | 60000 | 16-18 | 250000 |
| 1-30 | Silicone | 5000 | 2-3 | 30 |
| 30-300 | Epoxy Comp. | 9000 | 4-5 | 300** |
| 300-1400 | AMS Comp. | 16000 | 6-7 | 1000 |
| 1400-8000 | NVD Comp. | 20000 | 6-7 | 8000 |
| 8000-250K | Steel Comp. | 30000 | 6-8 | 250000 |

**Figure 9** Production volume versus tool type for injection molding a plastic interior cover for the sideview mirror of 95/96/97 Ford Contour and Mercury Mystique. *Conservative estimates based on experience. **The value can be increased using metal inserts in critical wear areas.

production volumes greater than 3000, the use of machined hard tools (tool steels) for injection molding becomes the most economical choice.

Using the RP&M approach, cast silicone is the economical mold of choice for making less than 30 parts. These molds can be used to cast polyurethane parts (a thermoset for which the durometer hardness can be adjusted to match polypropylene) that cost less and take 60% less time to make. Like the traditionally machined polypropylene prototype parts, cast polyurethane parts can only be used for simple mechanical tests to verify design intent. To make injection-molded polypropylene (prototype) parts that better represent production, the RP&M approach can also be used to greatly accelerate the fabrication of composite injection molds. Fabrication times can be reduced from 14 weeks to 6 weeks when compared to traditional machined tools. SL cross-linked photopolymer mold insert (i.e., Direct AIM) can be used to injection mold thermoplastics for low-volume applications. Care should be taken because SL material is brittle and is prone to premature fracture. For production volumes less than 1400 parts, composite tools with soft-shell surfaces can be used (up to 300 parts for CAFÉ or epoxy molds and up to 1400 parts for arc metal spray kirksite molds). For greater part volumes, composite tools with hard-shell surfaces can be used. In the tool-selection table, a 5000 part tool life was estimated for Ni-shell molds. Recent experience suggests that the tool life for Ni shells may be as high as steel-shell molds (up to 250,000 parts), making it the optiminal tool choice for high-volume injection molding.

## VII. SUMMARY

To remain competive in an evolving global economy, the automotive industry has pushed to institutionalize processes that provide speed to the marketplace. Recent trends have involved the reduction of product-development cycle time and manufacturing cost. In one way or another, these trends have involved the utilization of computer-aided fabrication technologies for accelerating the product and manufacturing development process. Efforts have been focused on developments involving automated fabrication technologies like RP&M for low-volume component manufacture that improve communication both internally and within the supplier base. Benchmarking Ford's injection-molded polyproplene interior cover (for the electric sideview mirror of the 95/96/97 Ford Contour and Mercury Mystique) as part of a lead-time reduction case study indicated a 55% reduction (from 30 weeks to 13 weeks) between traditional and RP&M. Replacing traditionally made patterns with RP&M-generated patterns not only reduces lead time but also allows for comprehensive design evaluations (formally a time-consuming iterative process) to be made early in the development stage for improved design quality in a fraction of the time normally required for traditional prototyping. In general, the application of RP to the product development process has shown a 60% decrease in lead time over traditional methods. Development efforts have involved (a) the machining of laminates (sheet or plate stock), where tool path generation can be totally automated, (b) RP, where complex 3D models or patterns can be automatically generated directly from CAD files without the use of molds or forming dies, (c) RT, where complex molds or dies can be made with minimal machining operations, and (d) RT&M, where RP, RT, and subsequent operations are applied to making parts with material properties close to or almost identical with the desired production material.

In time, future improvements in CAD visualization packages, together with the increased use of virtual reality, will likely enable designs to be more readily comprehended without the need for physical concept models. In addition, simulated finite-element trials can be performed on screen, and the increased use of CFD and FEA stress-analysis packages will reduce the need to perform flow and mechanical tests on physical components. Thus, the situation appears to be in a state of flux for rapid prototyping, with its predominant use today slowly declining over the next 20 years. This loss, however, will be more than compensated for by the major potential market which has yet to be fully exploited, namely rapid tooling.

Making tools for both prototype part development and production component manufacture represents one of the longest and most costly phases in the product-development process. The sequential approach to production tool fabrication by conventional machining is characterized by long lead times and high cost. Traditional tools (molds, dies, and fixtures) are machined from wrought tool steel billets. Tool steels like SAE 4340, H-13, and P-20 are most often used as die materials in production because of the unique properties obtained through alloying and heat treatment. Repair strategies that include welding is a must, to accommodate the frequent design changes that occur in the automotive industry. As a result, traditional tool-material selection is usually a compromise of properties (machinability and weldability versus wear and strength) affecting performance. This machinability compromise can be minimized using RT&M fabrication methods. The implementation of computer-aided technologies for improved communication between product and manufacturing will help eliminate these costly design changes by allowing for early problem detection in the initial stages of the product-development cycle. Additional ''upstream'' information flow between the manufacturing process and tooling fabrication stages encourages process-driven tool development for reduced fabrication lead time and cost (a rapidly growing future trend). Using RP&M technologies to accelerate the manufacture of tools will grow from its current economical use involving patterns for forming both soft prototype tools and bridge tools to the basis of making hard production tools directly from CAD data.

A variety of fabrication methods can be integrated with RP&M for future RT developments. For intermediate and high production volumes, RT advancements will be achieved by engineering a tool's physical and mechanical properties at the particle or molecular level. Technologies like 3D ink-jet printing, LOM, and SLS have been successfully modified to make RT directly from metal or ceramic powders. Indirect use of RP&M patterns for making PM molds have been applied by the Keltool process, PM casting, and PM forging. Powder materials electrolytically coated with organic binders are being investigated by a number of advanced material-development companies with some success. Companies like, Lone Peak Engineering, and Rapid Dynamics are using RT fabrication involving technologies. Making protective metal shells with cast aluminum, epoxy, or cement backing has shown significant promise for reducing lead time by over 30% compared to traditional tool fabrication methods. Unlike traditional tools, each component of these "composite mold assemblies" can be fabricated concurrently and assembled quickly to produce fully functional tooling. Because machining is minimal, tool life can be improved by selecting protective shell materials with greater wear resistance and heat-checking resistance than conventionally used tool steels. Several fabrication processes integrate well with RP&M technology and have shown promise for making protective metal shells: investment-cast steels, (arc/bulk) metal spray, NVD, electroformed Ni, and EDM. The idea of using optimal tool-selection tables for a part's RT technology application suggests future development of a multidimensional matrix for optimal tool selection (the process-driven "engineered" tool) for similar parts with respect to common design features, sizes, and materials.

The real challenge is not whether RT can meet niche-car market demands but whether RP&M can be effectively integrated into the automotive manufacturing-development process. Effective RP&M implementation will require large investments and will make extensive inventories of currently used capital equipment for material removal obsolete. This implementation will require fundamental cultural change in our product-development system. Our traditional system still relies on the use of 2D CAD part drawings and CNC machining. Most small automotive parts are outsourced to suppliers on a competitive basis. They have little incentive to invest, not to mention incorporate, these new technologies into their product-development process.

Among U.S. OEM automotive manufactures, over 67% of the average program investment is tooling related. Large part stampings like panels (body side, door, quarter, luggage), hoods, deck lids, fenders, and reinforcements account for 30% of the production tools and 70% of the cost. A future trend is to replace whole stamping assemblies with more complex plastic composites. Although consolidating a sheet metal assembly into a single more complex part takes longer, the total fabrication time and cost is far less than what is required to form and join simpler designed stamped components together. In addition, tooling requirements are much less severe than for traditionally stamped parts, making RT methodology more applicable. A development strategy is needed where U.S. OEM automotive manufacturers allocate resource efforts to RT for making large parts, leaving suppliers to focus their energies on implementing RT methodologies for making relatively small parts.

Rapid prototyping and manufacturing developments rely on solid CAD modeling and state-of-the-art computer-aided technologies. To date, only 8% of our design work force use solid CAD modeling and an even smaller percentage know enough about RP&M to take full advantage of its capabilities. The proliferation of RP service bureaus seem to have relieved the problem, but more capability and capacity will be required before RP&M can effectively supplement traditional NC machining for automotive manufacture. Cross-functional team efforts will be needed for improved communication both internally and within the supplier base while reducing technology development cost through multiple resource leveraging. In the 21st century, the automotive market will continue to become more global. To survive, U.S. OEMs must meet the demands of our competitive future by adapting a consumer-oriented mind-set on a global scale and become more flexible to change.

## REFERENCES

1. RJ Scheetz. Operating philosophy for low volume production. SME Report No. MF88-164, 1988.
2. LE Zeider. Automatic process generation and the SURROUND problem: solutions and applications. Manuf Rev 4(1):53–60, 1991.
3. LE Zeider, Y Hazony. Seamless design-to-manufacture (SDTM). J Manuf Syst 11(4):269–284, 1992.
4. T Sakuta. Development of an NC machining system for stamping dies by offset surface method. SME Report No. MS870657, 1987.
5. J Rohwedder. Urethane for sheet metal fabrication. SME Report No. MF750991, 1975.

6. S Wise. Net shape nickel ceramic tooling from RP models. SME Rapid Prototyping and Manufacturing Conference Proceedings, 1996.
7. Y Hazony. Seamless design-to-manufacture of marine propulsers: A case study for rapid response machining. J Manuf Syst 13(5):333–345, 1994.
8. S Schofield. Engineering research deployment teaching initiative: Reducing design-to-manufacturing time. Proceedings of the 1995 NSF Design and Manufacturing Grantees Conference, 1995.
9. J Dauvergne. Methodology and new tools for reduction of lead-time in product engineering example of HVAC development. SAE Internal Congress and Exposition, Detroit, MI, 1994; SAE Report No. 940885, SP-1035, 1994.
10. GR Glozer, JR Brevick. Laminate tooling for injection moulding. Proceedings of the Institution of Mechanical Engineers (IMecE), Part B: Journal of Engineering Manufacturing VQ07nb1 1993, p9-14 (0954-4054 PIBMEU), IMecE, 1993.
11. DF Walczyk DE Hardt. A new rapid tooling method for sheet metal forming dies. Proceedings of the Fifth International Conference on Rapid Prototyping, Dayton, OH, 1994, pp. 275–289.
12. DF Walczyk, NY Dolar. Bonding methods for laminated tooling. Solid Freeform Fabrication Symposium Proceedings, Austin, TX, 1997, pp. 211–221.
13. RC Soar, A Arthur, PM Dickens. Processing and application of rapid prototyped laminate production tooling. Proceedings of the 2nd National Conference of Developments in RP&T, Buckinghamshire, U.K., 1996, pp. 65–76.
14. FZ Shaikh, R Novak, J Schim, B Stroll. Precision stratiform machining: 100-day engine project. Prototyping Technology International '97, 1997, pp. 286–291.
15. T Nakagawa, M Kunieda, L Sheng-Dong. Laser cut sheet laminated forming dies by diffusion bonding. Proceedings of the 25th International Machine Tool Design and Research Conference, Birmingham, U.K., 1985, pp. 505–510.
16. P Dickens, D Sikon, R Sketch. Laminated tooling for moulding polyurethance parts. Proceedings of the SME Conference on Rapid Prototyping and Manufacturing, Dearborn, MI, 1996.
17. M Griffiths. Rapid prototyping options shrink development cost. Mod Plast 70(9):24–27, 1993.
18. S Odette. Complex assemblies from stereolithography and RTV tooling. SAE International Congress and Exposition, Detroit, MI, 1992; SAE Report No. 920744, 1992.
19. BA Jenkins. Epoxy tooling: Tomorrow's hopes are today's realities. SME Report No. EM860101, 1986. [Paper No EM 860101]
20. PF Jacobs. Recent advances in rapid tooling from stereolithography. Proceeding of the 2nd National Conference on Rapid Prototyping and Tooling Research. Buckingshire, U.K.: Buckingshire College, 1996. Mechanical Engineering Publications.
21. S Rahmati, PM Dickens. Stereolithography injection moulding tooling. 1997. Rapid Prototyping Journal v3n2 MCB Univer Press Ltd Bradford Engl p53-60 1365-2646 RPJORC.

22. E Sachs. Three dimensional printing: Rapid tooling and prototypes directly from a CAD model. Proceedings of the 1992 NSF Design and Manufacturing Systems, 1992.
23. CW Griffin, J Daufenbach, S McMillin. Desktop manufacturing: LOM vs pressing. Am Ceram Soc Bull 73(8):109–113, 1994.
24. S Michaels, EM Sachs, MJ Cima. Metal parts generation by three dimensional printing. Proceeding of the 4th International Conference on Rapid Prototyping and Manufacturing, 1993.
25. DL Bourell, RH Crawford, HL Marcus, JJ Beaman, JW Barlow. Selective laser sintering of metals. Manuf Sci Eng 68(2):519–528, 1994.
26. E Sachs, S Allen, H Guo, J Banos, M Cima, J Serdy, D Brancazio. Progress on tooling by 3D printing; conformal cooling, dimensional control, surface finish and hardness. Solid Freeform Fabrication Symposium Proceedings, Austin, TX, 1997, pp. 115–123.
27. K Gottschalk, V Cariapa, G Wick. Feasibility of stereolithography as an alternative to prototype patterns for high speed sand casting. AFS 99th Casting Congress, Kansas City, MO, 1995.
28. N Hopkinson, P Dickens. Thermal effects on accuracy in the 3D Keltool™ process. Solid Freeform Fabrication Symposium Proceedings, Austin, TX, 1997, pp. 267–274.
29. H Noguchi, T Nakagawa. Rapid tooling by powder casting transferred from RP model: Manufacturing conditions pursuing zero shrinkage. Solid Freeform Fabrication Symposium Proceedings, Austin, TX, 1997, pp. 287–294.
30. AT Anderson. Rapid tool fabrication by powder metal forging. SME Rapid Prototyping and Manufacturing Conference Proceedings, 1997; SME Report No. MF97–194.
31. DA Van Putte, LE Andre. A step-by-step evaluation of building an investment cast plastic injection mold. SME Rapid Prototyping and Manufacturing Conference Proceedings, 1995.
32. R Dzugan, RN Yancey. Investment cast tooling for metal casting and plastic injection applications using rapid prototyping. SME Rapid Prototyping and Manufacturing Conference Proceedings, 1997.
33. LE Wise, EL Grusoz, FB Prinz, PS Fussel, S Mahalingam, EP Patrick. A rapid manufacturing system based on stereolithography and thermal spraying. Manuf Rev 3(1):40, 1990.
34. R Gansert. Near-net shape manufacturing by plasma technology. Proceedings of the 1995 NSF Design and Manufacturing Grantees Conference, pp 423–424, 1995; SME Report No. B2382423.
35. A Mathews. Nickel vapor deposition tooling for the plastics industry. Proceedings of the Third International Conference on Advances in Polymer Processing, 1993.
36. JR Logsdon. Electroformed nickel tooling. SME Report No. TE880215, 1988.
37. P Jacobs. New frontiers in mold construction: high conductivity material and conformal cooling channels, SMR CMnCM 99–115, 1999.

# 10
# Rapid Tooling in the Medical Device Industry

**Daniel L. Anderson**
*DePuy Orthopaedics*
*Warsaw, Indiana*

> The significant problems we face cannot be solved at the same level of thinking we were at when we created them.
>
> —Albert Einstein

## I. INTRODUCTION

We have all heard the saying, ''problems are opportunities in disguise.'' Problems do, of course, present the opportunity to find solutions and, according to Einstein, require an entirely different level of thinking. Unfortunately, people often tend to look for solutions much more diligently when there is an urgent problem to be solved rather than simply planning ahead. Many facets of industry are currently faced with ''problems/opportunities'' in the forms of cost constraints, stiff competition, and reorganization of entire market segments. And, of course, once a competitor ''finds a better way,'' the better way soon becomes the new standard.

The health care industry is facing the same challenges: providing the best patient care possible while facing cost constraints from several different directions. Health care providers are forced to limit available funds and related services and/or share the expenses with patients, employers, the government, or private sources. Competing successfully in this global environment, where nationality, surgical expectations, and government regulation may dictate

**Fig. 1** Example of knee instrumentation for precision bone resection.

product requirements, is only possible as we embrace new technology and apply it to gain competitive advantage.

DePuy, a Johnson & Johnson Company, of Warsaw, Indiana, a leading orthopedic manufacturer, designs and manufactures replacement joints and implants for the musculoskeletal system of the human body, as well as related instrumentation (Figs. 1–7).

Degenerative and arthritic joint diseases often result in very painful and/or nonfunctional joint movement. Most commonly utilized are devices for the hip (Figs. 3 and 4) and the knee (as displayed in Fig. 5), where relief of pain and improved mobility are of primary concern. Products for other joints are also available, including prosthetics for the shoulder (Fig. 6), ankle, elbow, wrist, and so on. They consist of metal components typically made from a chrome–cobalt alloy, titanium, or stainless steel and are attached to prepared

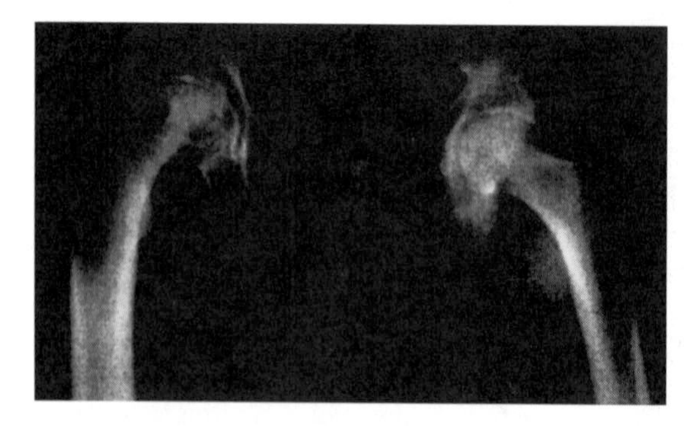

**Fig. 2** X-ray of degenerated hip joint.

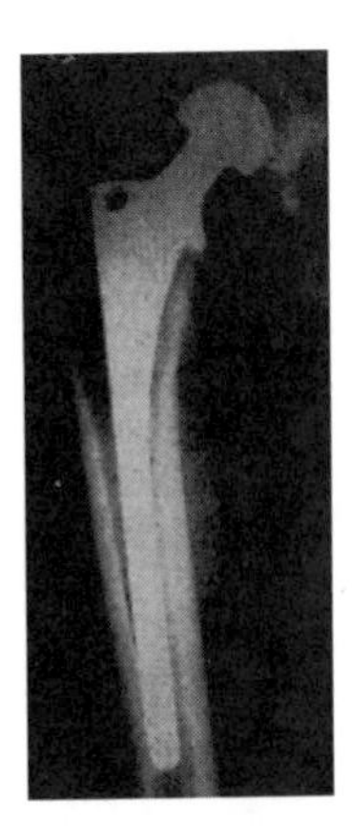

**Fig. 3** X-ray of hip replacement.

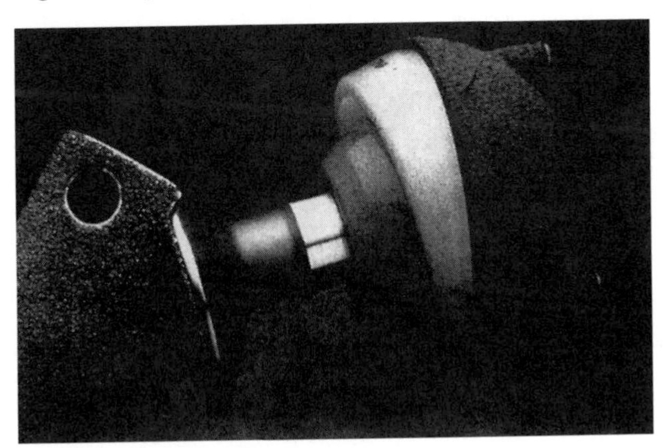

**Fig. 4** Hip stem with ball, cup, and liner.

**Fig. 5** Knee replacement system.

**Fig. 6** Components from shoulder replacement system.

bony structures and surfaces. Ultrahigh-molecular-weight polyethylene (UHMWP) bearings are often used between the mating joint surfaces. The components are often anatomically shaped or contoured designs, as opposed to basic geometric shapes (Fig. 7).

The implants are typically produced as a family in a range of sizes (Fig. 8). Also available are trauma, sports medicine, and spinal devices.

The medical industry has seen great advances in the quality of life offered to patient health care recipients. Many of these are related to various technologies such as imaging systems, laser scanning, robotics, and rapid prototyping and manufacturing technologies (RP&M) that are either coming of

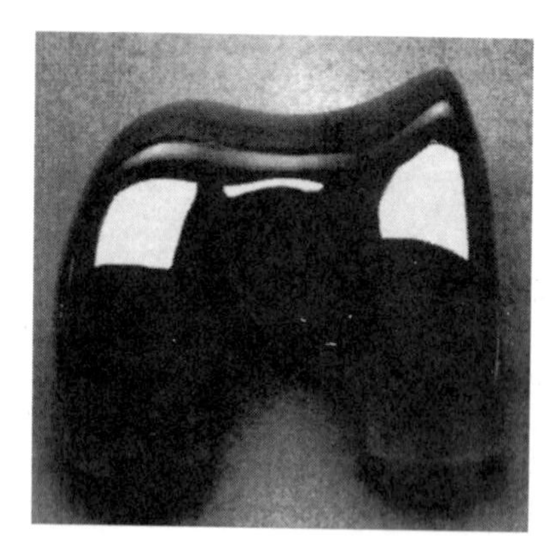

**Fig. 7** Complex geometry of femoral component for knee replacement system.

**Fig. 8** Stereolithography prototypes of five sizes of knee femoral components.

age or are now affordable for implementation. The medical industry, particularly orthopedics, has certain applications, which, although not necessarily unique, are extremely well suited to these technologies and may foster their integration and expansion. This chapter will discuss the ways that RP&M has enhanced the product development and the product launch phases of projects in the orthopedic field.

DePuy utilizes a series of RP&M systems that include stereolithography (SL) and fused deposition modeling (FDM).

## II. RAPID PROTOTYPING AND MANUFACTURING

Rapid prototyping (RP) is the term that has been coined for processes that can produce an accurate model from a computer-aided design (CAD) database without any additional tooling or machining. Because RP provides physical models so quickly, it has revolutionized the way many industries approach their product-development cycle(s).

Currently, other tools and technologies are being integrated with RP to produce capabilities that go far beyond the ''show and tell'' function of RP parts previously used for product development. RP models are being used as masters for cast tooling and sometimes to create the tooling or casting patterns directly. These capabilities, which are the basis of RP&M, are having a significant impact on industries throughout the world, including, of course, the medical industry.

Joint replacement manufacturers like DePuy are in no way immune to the health care upheavals of recent years. We have definitely had to ''change our level of thinking.'' As pressure from the competitive market and health-care-type reforms increase, we search for ways to design and produce products that are better and more cost-effective. We must work smarter. We are not simply competing for market share and position, but, rather, have at the very heart of our existence the desire to offer the latest and most technically capable devices to our customers.

## III. INVESTMENT CASTING AND CONVENTIONAL WAX PATTERN TOOLING

Investment casting can be traced back thousands of years to the ancient Egyptians and has been a staple of industry since. The basic process is quite simple: You make a pattern of what you want the end product to look like, coat it with a heat-resistant material to form a shell, melt or burn out the pattern, and pour in molten metal. The Egyptians sometimes hand-formed beeswax as patterns—a process that is probably well-suited for jewelry, but falls short for many other applications. If you want several dozen or several hundred metal parts and you want them identical, it becomes apparent that a process to mass produce patterns is necessary. The most common process is to make a mold—also referred to as wax pattern tooling—with the desired shape, and then inject wax into it to create wax patterns. (Perhaps some Egyptian jeweler came up with the idea of casting tooling when he had trouble hand-forming enough beeswax patterns to keep up with the demand of pyramid-shaped salt-and-pepper shakers or Sphinx bookends.)

We have all heard the saying ''garbage in, garbage out.'' The basic premise here is, of course, that what you get out of a process is never going to be any better than what you put into it. The same holds true for investment casting: If you want to produce great-looking, accurate metal parts, then you need great-looking accurate patterns. If you want to have great-looking, accurate patterns, you need great-looking, accurate tooling. Unfortunately, if you want to have great-looking, accurate tooling, you must understand that a very large percentage of the up-front cost, in both time and money, to get a casting program rolling will be to generate the pattern tooling. Most of us have had an experience where the cost of a casting was $30 per piece, but the cost for the tooling was over $15,000 and the lead time was 12 weeks. Also, if, for some reason, changes have to be made to the casting, you may find yourself going through the majority of the process again, with related costs.

So, if tooling is so costly to generate, why use investment casting? Well, there are basically four areas that must be considered: quantity, design, material, and speed.

*Quantity.* If you have to produce a large number of parts, then investment casting is often a very cost-effective mass-production method. The up-front cost of the wax pattern tooling is nicely amortized. However, if only a small number of parts are needed, it generally is not good business to dump a large amount of money into tooling; your per-piece cost will probably be disproportionately high.

*Design.* Generally speaking, the more complex the design, the more machine and/or assembly time will be required to produce the product. Fabrication of the complex shapes required for orthopedic joint replacements would require many hours of surface machining. Investment casting can often be a cost-effective method to produce complex parts—even for a relatively low number of parts—if, of course, the up-front cost of the tooling can be offset relative to the cost of the alternative. Again, if the product is complex, the tooling will usually be complex—and expensive.

*Material.* Some materials are much more difficult to machine than others. For example, cobalt–chrome polishes nicely and interacts well with UHMWP bearings, but its material properties make is less than pleasant to machine. Again, investment casting can provide some relief, if you can design castings such that there is little finishing work required to produce the end product.

*Speed.* Simply put, sometimes you can live with the lead times required to develop wax pattern tooling, and other times you cannot.

Most manufacturing situations require a combination of these factors to reach a satisfactory production decision. For example, the vast majority of DePuy products that are made via investment casting are cobalt chrome and consist of geometric shapes that would require extensive material removal if machined. But the quantities vary. Sometimes, we need a small number of implants for a clinical study; or, in the case of a patient-specific or custom implant, we may need just one.

# IV. CONVENTIONAL TOOLING MANUFACTURE VERSUS RAPID-TOOLING MANUFACTURE

If tooling were substantially less expensive and faster, or if there were a way to produce accurate patterns quickly and cost-effectively without tooling, what would be the impact on industry? Or more specifically, DePuy?

- Low-quantity casting runs could be more readily utilized for custom implants, regional products, and clinical studies.
- Lower overall casting costs could increase profit margins, reduce the cost of the end product to the customer, or both.
- Functional first-article castings could be obtained much faster for debugging finishing operations and/or to speed up product launches.

## A. Some Tooling Alternatives

Several years ago, during a situation in which we needed to launch a product quickly, the question was asked, ''If we can make prototype parts with an SLA machine, why can't we make wax pattern tooling?'' Good question. Consequently, we designed the wax pattern tooling on our CAD system and built it with a SLA. (Figure 9 shows core and cavity done in stereolithography (SL), on the left.) There were some inherent problems with the approach, but it served its purpose and the principle was established: We could design and launch a product with wax pattern tooling that we had quickly generated ourselves (i.e., rapid tooling).

As mentioned, there were some problems with the approach of building wax pattern tooling on a SLA. The cured photopolymer was brittle and several

**Fig. 9** Examples of SL (left) and epoxy tooling for investment casting.

three-piece tools ended up being ''too-many-pieces-to-count'' tools during the wax-injection process, and there was some ''stair-stepping'' on angled surfaces that was difficult to smooth out in the internal areas. We started looking at other possibilities and eventually reached the conclusion that we would be better off if we made an SL model, smoothed out the surfaces, and formed a material around it to produce the tooling (Fig. 9, core and cavity on the right). We used this method for a few years for product launches and products where we would run several wax patterns, store the tooling, and, in a few weeks or months, run more patterns. Unfortunately, we found that the material that we were using for the tooling was susceptible to moisture, and after being stored, it was no longer producing accurate patterns because of the resulting dimensional instability.

Nonetheless, we knew that we were on the right track; it was just a matter of finding the right material. We eventually purchased a spray-metal system that coats the master with a metal surface. We pour a mixture of aluminum beads and epoxy resin on the back of the metal coating to finish the tooling. (An example of a spray metal tool is shown in Fig. 15.) More details on this process will be included in the hip stem case study.

## B. Direct Pattern Generation

Another tooling alternative is no tooling at all! Also known as direct pattern generation, this is a wonderful alternative to creating wax pattern tooling for small-to-medium quantity casting runs. This is a very welcome development and has several distinct advantages, the most obvious, of course, being the fact that you will not incur *any* expenses related to tooling. Another advantage is the ability to tackle projects that are cost-prohibitive when considering traditional methods (more on this in the custom knee case study).

The basic process consists of an RP system that produces parts (patterns) that can be used directly in the investment-casting process—being burned out of the ceramic shell, completely bypassing the need for wax pattern tooling. Nearly all of the RP companies are now offering some form of direct-pattern-generation process. Many of the companies use thermoplastic materials (materials that will melt) in their machines, so any part that is generated is a potential casting pattern.

DePuy is using 3D Systems' QuickCast™ process. We first started experimenting with direct pattern generation in 1992, before QuickCast was introduced. At that time, there was a resin that was supposed to be ''investment castable.'' The patterns that we produced were solid—like all of the other prototype parts that we made—instead of the honeycomb-like internal structure of QuickCast. The resin we were using was a thermoset material (material that will not melt) and there were some problems with shells cracking during the "melt-out" phase of the investment-casting process, resulting in the possibility of corruption in the castings.

## V. CASE STUDIES

There are basically two categories where we use QuickCast: product launches and custom implants.

### A. Hip Stem Case Study—Product Launch

The following case study of a hip implant system shows how rapid tooling impacts the product-launch process. The entire system includes 28 stems with various neck angles and sizes, consisting of both right and left designs. The goal here was to produce several of the intermediate sizes in order to facilitate a clinical launch of the stem design. A clinical launch refers to the implantation of several stems by the designing surgeons, in order to get a "real-life" feel for the product and perhaps suggest last-minute improvements. Consequently, design changes evolving from the study were a very distinct possibility.

There are three basic reasons why we took the rapid tooling approach in the clinical launch of this product:

1. Reduced cost
2. Reduced lead time for tooling
3. Changes from information obtained during the clinical period would probably result in modifications to the product

Cost is everything. Well, okay, so it is not everything. But nearly every consideration can be traced to cost concerns: cost of the product (of course), cost of market share and sales lost because of delayed launch, and cost of selling a product that did not have that one last opportunity for improvements. Each of these cost issues are related to the three reasons listed.

Let us take a look at a hypothetical example of how RP&M impacts the amount of time needed to launch a single hip implant. In this example, the lead time for the conventional machined wax pattern tool is 12 weeks. After the completion of the tool, it will take another week to get castings, and another 2 weeks for postprocessing. Furthermore, the time necessary to rework the tool for minor design changes is typically about 4 weeks.

Given this information (and simple arithmetic), it can be seen that it will be 13 weeks before the first castings are completed and, consequently, that long before manufacturing even gets to have a look at them. The customer gets access to the product after 15 weeks, and if the customer or the manufacturing team wants to make changes, the time needed for tool rework kicks in. At that point, the decision must be made whether to continue production with the existing castings and phase in the changes, or completely halt production until the ''new'' castings are available. If the decision is made to wait on the new castings, the finished product would not be available for 19 weeks or about 4.5 months! The key point is that the interval needed to generate and/or rework a tool will almost always add large chunks of time to delay the release of a product. Although this example is hypothetical, it is not too far off from the actual case study.

In an ideal situation, tooling would be generated in a few days and castings would be available in 2–3 weeks, allowing Manufacturing an earlier opportunity to begin working with them. Also, the modus operandi would be inexpensive enough to allow design ''improvements'' without turning everybody gray. This is exactly what RP&M offers (but with no guarantees against going gray). Another added perk is the confidence of seeing the first wave of the new design *before* committing to spending significant funds for production tooling. (This ''reduced commitment'' is often a motivation for the expedient approval of a design.)

## B. The Situation

The designing surgeons believed that this hip replacement system was an excellent product and they wanted to have access to the implants as soon as possible. We believed that this hip replacement system was an excellent product and we wanted the surgeons to have access to the implants as soon as possible. The designing surgeons like to have some flexibility to make design changes, based on the knowledge that they gain during the first several implant surgeries. We also like the surgeons to have some flexibility to make design changes, based on the knowledge that they gain during the first several implant surgeries. However, we want to be able to launch a product in a period of time that matches business objectives.

The manufacturing group also likes to have some flexibility. As seen in Fig. 10, this implant required considerable postprocessing. Based on the knowledge that Manufacturing personnel gain from the first several castings, changes to the design are possible to improve manufacturability and reduce the final cost of the product.

Thus, the issues basically boiled down to two: we need implants fast and we need the flexibility to make improvements without having to throw

**Fig. 10** Multiple views of completed hip stern.

expensive and time-consuming hard tooling into the recycle bin. By the time the first several castings had been implanted, we had good feedback on possible design ''tweaks'' from Manufacturing and the surgeons. It is clear that rapid tooling increases your ability to take risks, as the tooling is neither prohibitively costly nor likely to seriously delay a product launch.

## VI. THE RAPID-TOOLING PROCESS

The development process does not drastically change simply because rapid tooling is used rather than conventional tooling methods. The implant still must go through its design phase, prototypes must be generated, and so forth. Where the greatest impact takes place is in the initial launch of the product and the decisions of when (or in some cases, ''if'') it is appropriate to replace the rapid tooling with production tooling.

The actual process to create the tooling is really rather straightforward, but not as seamless as one may like. For instance, it would be great if all you had to do was design the implant, produce an SL model of it, and pour a mold around it. However, you find out quickly that you must start dealing with such issues as pattern and metal shrinkage, sacrificial material, gates, and so forth. Even these issues are not terribly burdensome if the proper approach is taken.

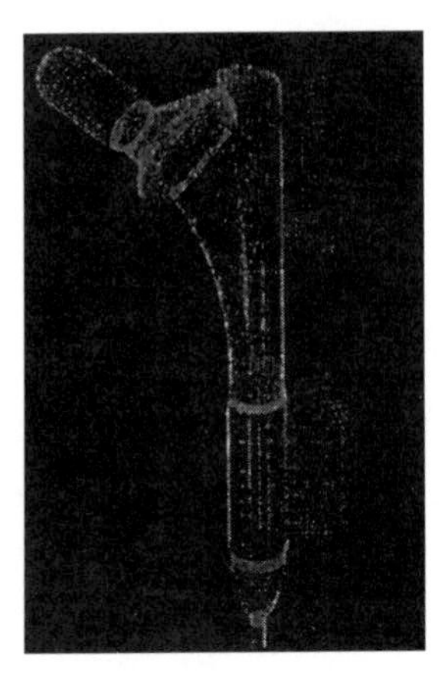

**Fig. 11** Solid CAD model of casting design for the hip stern.

The process usually has the following steps:

- Design the implant
- Produce a CAD model of the implant
- Design the casting pattern
- Produce a CAD model of the pattern (Fig. 11)
- Provide shrinkage compensation to the pattern
- Produce an SL model of the pattern (Fig. 12)

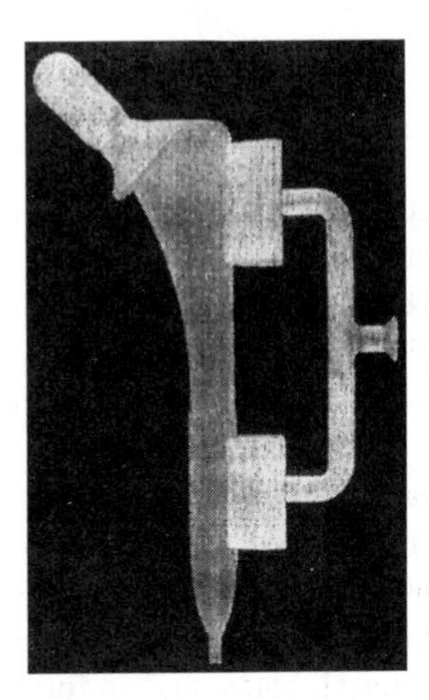

**Fig. 12** Stereolithography model of casting design of the hip stern.

**Fig. 13** Mounting of SL model for tooling fabrication.

- Inspect the pattern
- Produce the tooling using the SL pattern as a master (Figs. 13–15)

Again, as you compare the initial metal casting with the finished product (Figs. 16 and 17), you can see the amount of postprocessing that is necessary.

## A. Knee Implant Case Study—Custom Implants

The following case study for a custom femoral knee implant shows how direct pattern generation can impact the production process. The major goal is to reduce both cost and time to a level where a custom knee implant can be generated cost-effectively and without negatively impacting other projects.

If the only option were to create traditional machined wax pattern tooling, then the cost of the project would be prohibitive, as the level of complexity of the impact design would necessitate a four-piece wax pattern tool. One

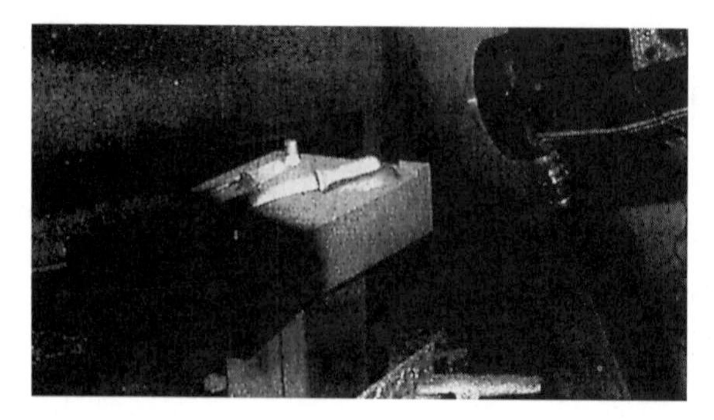

**Fig. 14** Applying metal coating to create one-half of the tool.

**Fig. 15** Completed tool with wax pattern.

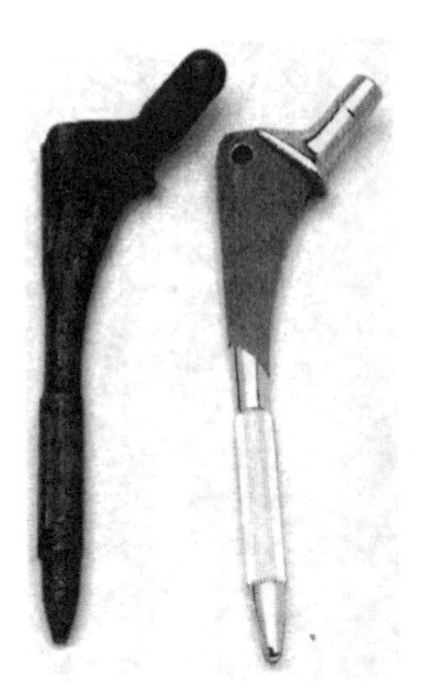

**Fig. 16** Raw casting and completed hip stern.

**Fig. 17** Completed hip stem with mating components.

alternative to complete this project is to machine patterns from wax, thus avoiding the cost of the complex tooling while providing patterns. This is a solution that we have used in the past, but it is more labor-intensive and, ultimately, more costly, in most cases, than acceptable for a custom implant. Another alternative would be to generate the implant by surface machining it from CoCr. Again, this is a labor-intensive, costly choice compared to acceptable standards.

Direct pattern generation seems to be tailor-made for this type of scenario. There are no costs or lead times associated with generating wax pattern tooling and no machining is required to produce wax patterns or, even more costly, the implant itself. Also, in the event of a late design change, you can react much more quickly than with conventional methods.

In this case, the goal was to produce a revisional femoral knee implant that would compensate for the patient's bone loss in his distal femur while providing the highest possible level of functionality and integration with the existing tibial components.

The information available consisted of x-rays and communication with the surgeon. The patient had been through a series of knee surgeries since the late 1970s and now had insufficient bone to enable use of an off-the-shelf femoral implant of the size required to maintain joint functionality. Consequently, the latest implant was sized to fit the remaining bone, but was too small for proper joint function (Figs. 18 and 19).

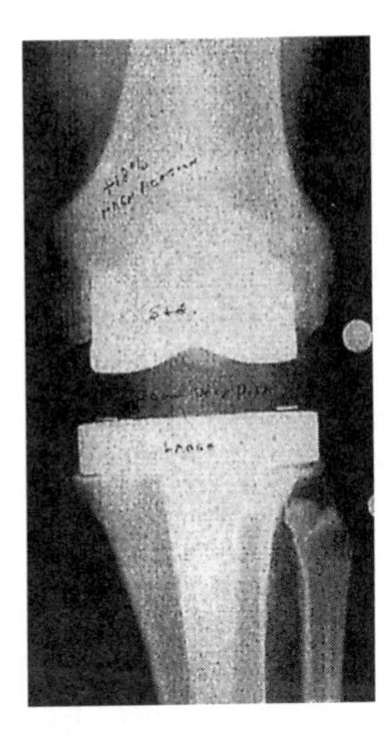

**Fig. 18** X-ray of subject's knee joint—front view.

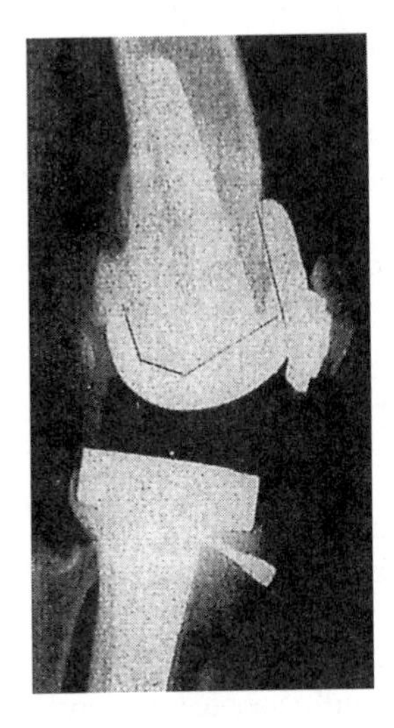

**Fig. 19** X-ray of subject's knee joint—side view.

The design of the custom femoral implant would have to consist of an articular surface large enough to maintain joint integrity, match the existing LCS (i.e., ''Low Contact Stress,'' the trade name of a very successful knee system developed and marketed by DePuy), the large revisional tibial component, and still be small enough to make up for the loss of bone.

It was decided that the design should consist of an articular surface matching an LCS size Large, and also matching a smaller implant in the LCS line—with a stem added for increased stability. Because this design incorporated existing shapes from the LCS system, it simplified the entire design process and eliminated the need for custom instrumentation.

Once the design was established, we were faced with several options for the actual production of the implant. If we limited ourselves to traditional production methods, the cost of the product could be prohibitive. Because cobalt chrome is the material of choice for femoral knee implants, the cost to machine the implant would be high because of the properties of this material. A near-net casting would be ideal. However, this would have to include tooling to produce the casting wax patterns; tooling that would be very complex and probably necessitate a four-piece design, and, again, would be very costly. Another alternative to complete this project is to machine wax patterns for investment casting, avoiding the expense of machining cobalt chrome and the cost of complex tooling. This is a solution that we have used in the past, but it is more labor-intensive and generally more costly than considered cost-effective for a custom implant.

Direct casting pattern generation via QuickCast on a SLA provides an excellent solution. There are no costs or lead times associated with generating casting tooling, and no machining required to produce wax patterns or, even more costly, the implant itself. Also, in the event of a late design change, you can react much more quickly than with the traditional methods. The process consists of creating a model on the SLA that has a structured interior but is mostly hollow. The QuickCast model is then used in place of the wax pattern and will collapse, not expand, during the burn-out phase of the investment casting process.

Basically, the process consists of these (familiar) steps:

- Design the implant
- Design the casting needed to produce the implant
- Design the pattern needed to produce the casting
- Generate a CAD model of the pattern (Fig. 20)
- Generate the pattern via QuickCast (Fig. 21, left)
- Generate a casting from the pattern that will, hopefully, closely match the casting design (Fig. 21, center)

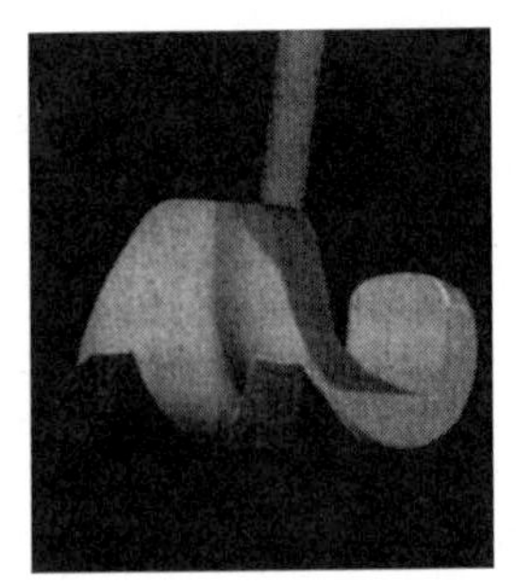

**Fig. 20** Solid CAD model of casting design for the knee component.

- Finish the casting to match the implant design [Fig. 21 (right) and Fig. 22]

As you can see, wax pattern tooling, the costliest step, is completely absent from this procedure. Regardless of the method of production, the first two steps will have to be completed. If the production method were to generate alternative tooling or machine the wax patterns, it would still be necessary to do the first three steps.

A solid model of the implant was designed in CAD using the established design criteria. A solid model of the implant casting was then created, adding material for polishing and finishing. The casting pattern was then created in CAD by scaling the model to compensate for the shrinkage of the cobalt chrome and, finally, adding gates. The latter information was obtained from the casting vendor. The next step was to generate QuickCast models of the casting patterns via an SLA. The QuickCast models were sent to the casting

**Fig. 21** Left-to-right: QuickCast pattern, raw casting, finished product.

**Fig. 22** Closeup of finished product.

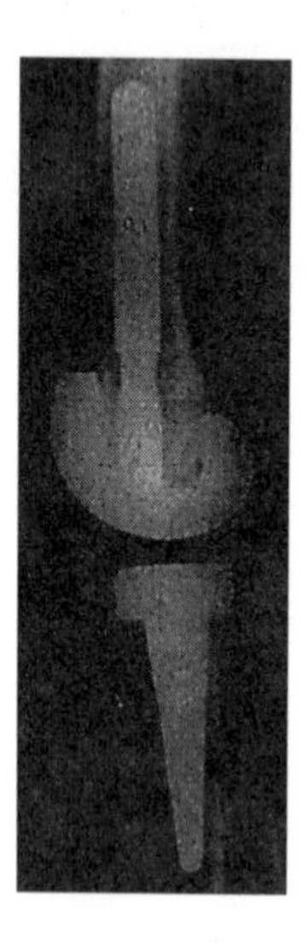

**Fig. 23** Postoperative x-ray of subject's knee joint—side view.

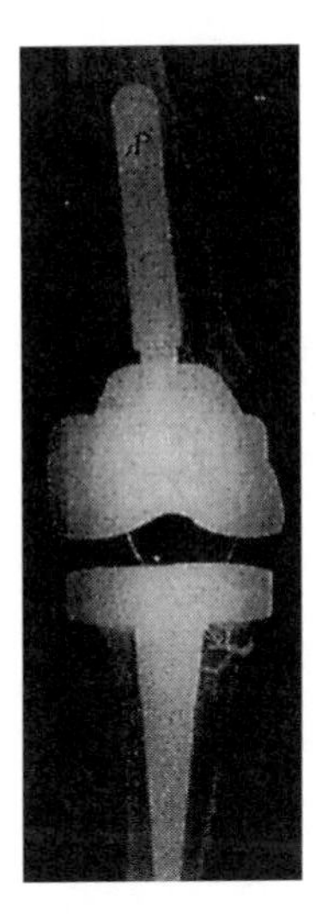

**Fig. 24** Postoperative x-ray of subject's knee joint—front view.

vendor, and the castings were received about 2 weeks later. The castings were then finished, with one to be used as the implant and one as the trial.

The surgery took place early in 1995; the patient has since displayed the results of a very successful surgery. The implant, by design, did not require any revision of the tibial component. The patient has the added benefit of a one-piece femoral implant rather than one consisting of multiple components required to fill in the areas of bone loss (Figs. 23 and 24).

## VII. CONCLUSIONS

These case studies are examples of how RP&M technologies make possible the creation of products that can improve lives through enhanced preoperative planning and custom implants. As the use of RP&M and especially rapid tooling becomes common in the medical industry, more and more applications of this valuable technology will be discovered and utilized.

## ACKNOWLEDGMENT

All photographs were provided courtesy of DePuy Orthopaedics, Inc.

# 11

# The Role of Rapid Tooling in Investment-Casting Applications

**Thomas R. Richards**
*American Industrial Casting, Inc.*
*East Greenwich, Rhode Island*

**Hugo Lorrain**
*Howmet Aluminum*
*Laval, Quebec, Canada*

**Peter D. Hilton**
*Technology Strategies Group*
*Concord, Massachusetts*

## I. INTRODUCTION

Rapid prototyping (RP) and investment-casting technologies have been used in conjunction with one another since the early 1990s for various purposes. One use of RP in support of investment casting has been to form the pattern.

Early work was with wax patterns produced by selective laser sintering (DTM Corporation) or ink-jet printing (Sanders Prototype, Inc.). An alternative, involving plastic patterns produced by stereolithography (3D Systems), was attempted and subsequently refined. Early patterns caused cracking in the ceramic shell when they were burned out. Producing patterns with a continuous surface and a honeycomb interior (3D Systems QuickCast™) solved this problem.

Today, RP technologies are used in conjunction with investment casting for at least two purposes: (a) to produce patterns for use in manufacturing prototype parts or very small sample sizes and (b) to produce patterns for tooling (molds) which can then be investment cast. The tools are then used to produce investment-casting wax patterns or for other mold-related processes such as injection molding. The uses of RP in conjunction with investment casting are relatively mature and in commercial application at numerous casting firms.

This chapter contains descriptions of the use of RP in investment casting at two firms. The first, American Industrial Casting, Inc., is a relatively small and very innovative company. It is an example of a lead user among smaller U.S. firms and demonstrates cost-effective implementation with relatively modest capital investment. American Industrial Casting provides investment castings to several market sectors. Its parts are typically small but detailed, with tight tolerance requirements. The second firm is Cercast, a division of Howmet, which is one of the largest producers of investment-cast aerospace components. Cercast uses RP to form patterns for prototype investment casting of large, complex aerospace parts.

Tom Richards is the technology leader at American Industrial Casting, Inc. Hugo Lorrain is responsible for prototype investment casting at Cercast. They each describe some results achieved by their respective companies.

## II. RAPID TOOL MAKING FOR INVESTMENT CASTING AT AMERICAN INDUSTRIAL CASTING, INC.

American Industrial Casting, Inc. (AIC) is a manufacturer of precision investment-cast parts. It produces castings for the aerospace–defense, electronics–communications, mechanical components, medical, and subminiature parts industries.

These parts are characterized by the relatively small size (typical part dimensions usually fall within a 7-in. cube in solid molds, although parts to 24 in. are produced in shell molds). Their requirements include fine-feature definition of the order of 0.003 in. radii, walls as thin as 0.011 in. with high aspect ratios, and tolerances of ±0.003 in./in. up to 0.5 in. and ±0.005 in./in. above 0.5 in.

American Industrial Casting, Inc. focuses on producing production quantities of finished parts in nonferrous or ferrous alloys from hard tooling produced in 10–12 weeks. These parts are often intricate in detail and thin walled. AIC assists customers in their design for manufacture and assembly efforts by providing rapid prototyping and manufacturing (RP&M) generated castings.

Using the traditional RP&M approach (i.e., building RP&M patterns as positives of the final part), AIC's development cycles are typically 1 week for nonferrous alloys (using solid molds), or 3 weeks for ferrous alloys (using shell molds). These times are from receipt of the customer's three-dimensional (3D) solid model transferred in .stl file format. Because the RP patterns are destroyed by the lost-wax investment-casting process, this approach is most cost-effective when a customer is buying only a few parts.

In response to customers' demands for the rapid prototyping of more than a few parts and first production, AIC became involved with a number of rapid-prototyping technologies. Beginning in 1992, AIC used service bureaus to produce rapid prototyped patterns from customers' solid models. This experience gave AIC an initial sense for the capabilities and limitations of the various RP&M technologies. In general, the then available technologies were not able to achieve the tolerances and surface finishes required for AIC's applications. Further, the RP materials were not optimal for investment-casting applications. Beginning in 1994, AIC carried out a systematic study of the various RP&M systems available on the market in terms of their particular requirements. AIC chose to buy the Model Maker™ System from Sanders Prototype, Inc., of Wilton, New Hampshire, as that most closely matching its needs in terms of part size, resolution, surface finish, and tolerance capabilities, as well as material properties.

The Model Maker System is able to achieve fine-feature resolution and to produce thermoplastic prototypes that can serve as casting patterns. This RP&M system uses ink-jet printing techniques to lay down droplets of resin and wax from separate injection heads, one layer at a time. The prototype geometry is created in the resin and the remainder of the space is filled with the wax. In this manner, the wax supports any down-facing surfaces and fills internal cavities during construction. Upon completion, the wax, which has a lower melting point, is melted away, leaving the freestanding thermoplastic as a pattern suitable for investment casting.

American Industrial Casting, Inc.'s initial concept was to use the RP&M system to produce investment-casting patterns and to use the patterns to produce investment castings by the traditional lost-wax route. The advantage was that a prototype casting could be developed in days. The disadvantage was that an RP pattern is needed for each casting, and, consequently, this approach becomes both uneconomical and slow for batches of more than several investment-cast parts.

American Industrial Casting, Inc.'s first approach for producing more than several prototype parts rapidly and cost-effectively was to use the RP pattern (appropriately sized to account for shrinkages) as a pattern for casting a beryllium–copper alloy master. This metal master was then used to produce temporary molds of vulcanized rubber, room-temperature vulcanized polymer (RTV), or epoxy. These molds were, in turn, used to create wax investment-casting patterns for producing parts. Unfortunately, these transfer-molding methods resulted in problems. In the order mentioned, dimensional variations ranged from ±0.060 in. to ±0.015 in. to ±0.005 in., geometric distortions from severe to moderate to marginally acceptable, mold-building times from hours to weeks to months, and costs ranged from $50 to $500 to $2500 or more.

American Industrial Casting, Inc. set out to develop a different method: one in which RP castings are produced for the injection mold components from RP patterns. The approach is straightforward. The designer, starting with his 3D computer-aided design (CAD) final part design, is coached in the creation of ''shells'' around his part. Parting planes are installed in such a way that the shells can be removed from around the injected wax or plastic pattern without being damaged. These individual tooling components are built as solid objects, molded by AIC's solid-mold process and cast in a beryllium–copper alloy. The process takes only 2 days. The alloy is very fluid and duplicates every feature of the pattern down to the finest detail and finish. The resultant metal mold components are assembled using conventional mold finishing techniques. Waxes are then injection molded. The waxes are then assembled, either into solid molds for nonferrous castings or into shell molds for ferrous castings. Solid molds are produced by pouring a slip of refractory investment material around the wax patterns in a vacuum environment and allowing the slip to solidify within several minutes into a solid mold, which is dried and fired overnight and held at a suitable temperature for pouring the next day. Shell molds are formed by dipping the assembled waxes into a slip of refractory investment material and hanging up to dry under controlled conditions, successively adding layers of ceramic, over a several-week period, to complete the shell mold, which is fired prior to casting. The benefits of this approach are lower cost and more rapid development of *injection* molds capable of producing reasonable runs of hundreds of functional metal parts. The molds are actually capable of producing hundreds of thousands of parts, limited only by the high cost of hand injection molding. AIC's ultimate goal is to use their process to make and install die cavity insets into rapidly produced and economical machine injection molds that are suitable for the production of hundreds of thousands of waxes for producing precision investment-cast parts. In short,

AIC is using investment casting to make wax pattern tooling for investment casting!

These benefits result from the direct transfer of 3D CAD geometry to a physical geometry, followed by the use of standard investment-casting technology to produce metal alloy cast parts as-designed. The use of RP&M patterns plus investment casting of the mold components substitutes for the more traditional computer numerically controlled (CNC) machining of the molds, which typically requires 10 weeks.

Given these concepts, we need to quantify the capabilities and limitations of the RP&M direct tooling process as applied by AIC. The Sanders Model Maker System defines many of these capabilities and constraints. The current model MM6B Model Maker Pro's are able to hold in-plane tolerances of ±0.001 in. The Z-direction resolution is set by the layer height. Layer heights can be selected between 0.005 in. (coarse resolution) and 0.0005 in. (fine resolution) with corresponding impacts on build time. Build times involve 28 s per layer of fixed time, plus build rates that vary from 0.02 to 0.40 in.$^3$/h, for 0.0005–0.005-in. layering, respectively. As an example, the cavity for an intricate part about 1.5 × 1.5 × 0.75 in. might be contained within mold halves each measuring 2 × 2 × 0.5 in. overall. If 0.002-in. slicing were selected, a build rate of 0.18 cubic in.$^3$/h could be expected, plus a fixed time of about 28 s per layer for milling. Thus, the two mold halves could be built in about 24 h. Add 48 h for the solid mold process and an injection mold can be ready for assembly and finishing in only 3 days! Although comparatively slow, the process runs unattended overnight and builds patterns that are accurate, resolute, and smooth (80–100 RMS) on all surfaces. Patterns up to 6 in. can be built on the Sanders MM6B. Sanders Prototyping Inc. is continuing development of RP systems based on its technology as well as technology refinements, so we expect that, when you read this material, their machine's capabilities will have been improved from the numbers given here. AIC also employs service bureaus using other RP processes for economical building of larger patterns.

An example part produced by AIC using the ''Prototype the Tool'' process is shown in Fig. 1. The part shown resulted from a CAD model of a diode box which had just been put into production at AIC using conventionally CNC machined tooling. It was chosen as a fair challenge for a first RP&M tooling demonstration project.

Virtual Concepts Design at virtcon.com was engaged to produce both a 3D CAD model of the part and of the RP&M tool components for the diode box working from the customer's 2D print. Figure 2 shows the diode-box injection-mold components placed alongside one another for the building of

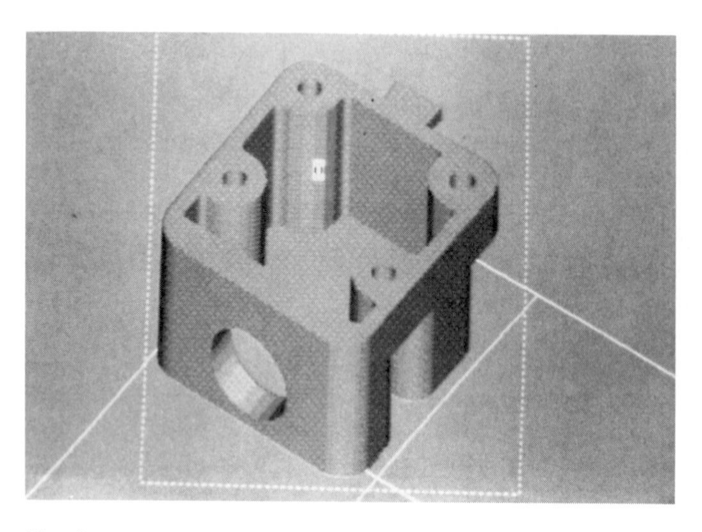

**Fig. 1** An example part produced by AIC using the ''Prototype the Tool'' process.

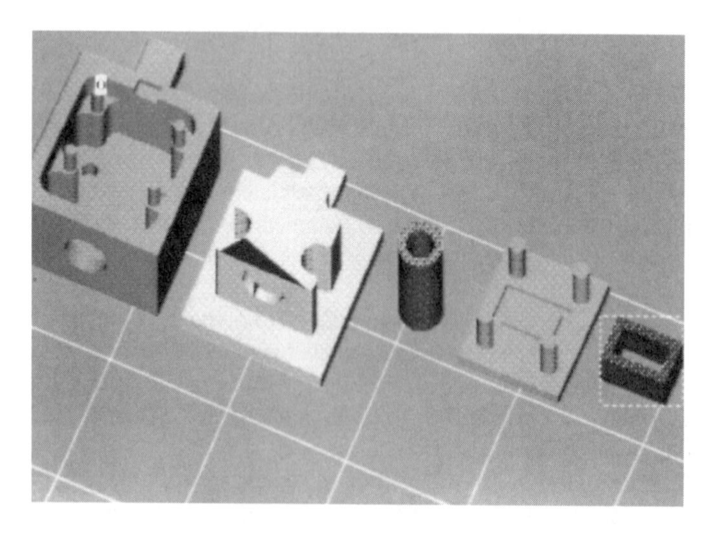

**Fig. 2** Diode-box injection-mold components placed alongside one another for the building of the patterns.

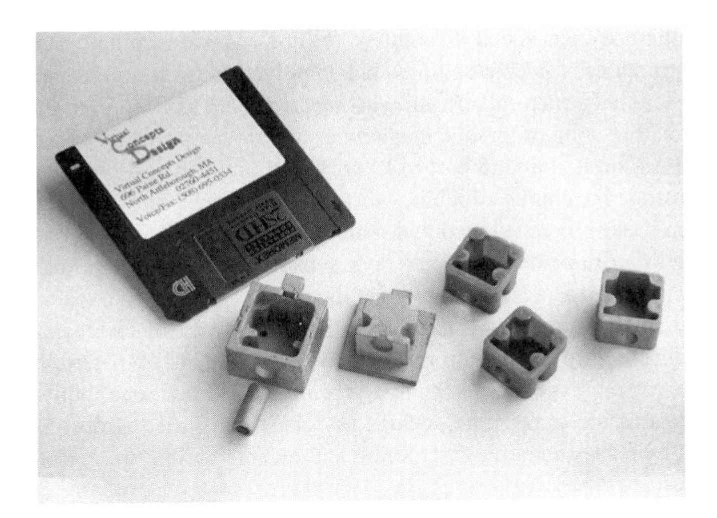


**Fig. 3** Floppy disk containing the data for the mold components (left), the final cast mold components (center), two wax patterns produced from the tool (right center), and a final investment-cast part in aluminum A356 alloy (far right).

the patterns. Figure 3 shows the floppy disk containing the data for the mold components (left), the final cast-mold components (center), two wax patterns produced from the tool (right center), and a final investment-cast part in aluminum A356 alloy (far right).

Once the files were made available, the entire process producing the RP&M tool required only 1 week. The first 12 castings required another 2 days plus an additional day for heat treatment. Consequently, only 10 days after receipt of part files, 12 functional, heat-treated, metal castings were available for the customer. Only a few years ago, this would not have been possible.

## III. RAPID PROTOTYPING, THE MODERN TOOL FOR DEVELOPING CASTING APPLICATIONS AT CERCAST

Thin-wall, dimensional, high-strength aluminum investment castings have gained significant visibility in the past few years, as a highly credible method of producing demanding airframe components. The production technology has replaced (a) multipiece sheet-metal assemblies, (b) hogouts and forgings, and (c) composite structures, for cost savings and improved damage tolerance. Although the benefits of structural airframe castings are being realized in most new fixed and rotating wing programs, the number of new sample and tooling programs have been limited due in part to lengthy lead times. Rapid prototyping has demonstrated a unique capability to provide certification hardware while production tooling is developed and matured in a parallel effort.

According to airframe designers, ''a door substructure (such as the one in Fig. 4) can be designed as a precision casting much faster than an equivalent multipiece sheet metal fabrication.'' Part count reduction, lack of fasteners, absence of custom shim stock, and elimination of an extensive bill of materials make a complex casting easier to design and procure than a traditional built-up structure. Unfortunately, the time savings associated with casting design is usually offset by subsequent lengthy tooling and manufacturing times. For

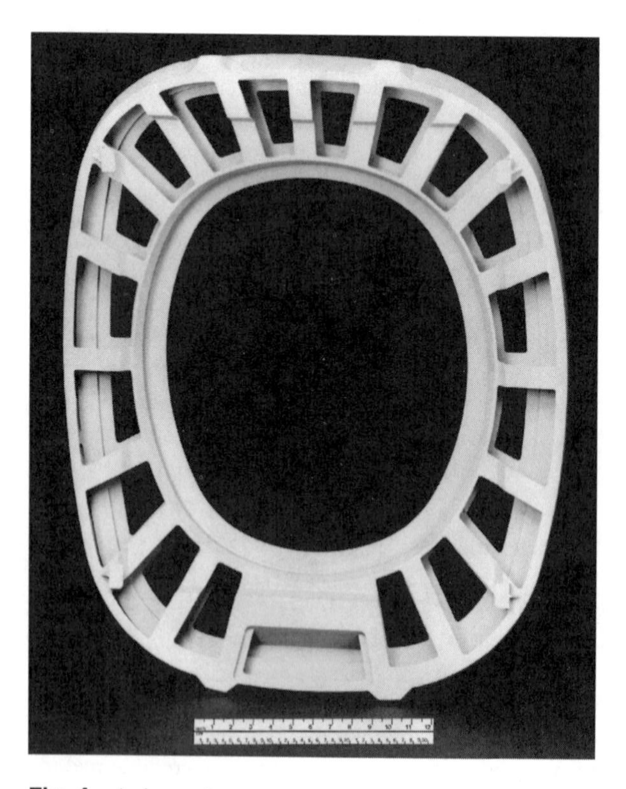

**Fig. 4** A door substructure.

programs with adequate lead times of several months, structural castings are highly competitive and often specified. Rush retrofit or ''short fuse'' development programs, however, often pass on advanced casting technology, except where rapid prototypes can be obtained. In many programs, the ability to procure rapid prototypes is the deciding factor to design, test, and certify castings into a production program.

There are no theoretical size limits to the RP process, and structures can be cast with similar wall thicknesses, strength, and size scale as production casting hardware. Castings excel at delivering components of high complexity, incorporating many ''next assemblies'' into one single component. Reduction of machining, joining, and tolerance stack-up from multicomponents provides for unique structures.

Concurrent industry developments in recent years have yielded larger, more accurate, and smoother RP patterns, in addition to reliable casting technology to transform these patterns into metal hardware. A description of the complete investment-casting process can be found in numerous literature sources. The RP process substitutes a pattern quickly produced using stereolithography for the heat-disposable wax/polymer pattern normally produced from a production injection tool. In bypassing costly and time-consuming tooling, the foundry engineer can use this RP pattern to form a precise ceramic mold, followed by pattern removal, mold curing, and subsequent casting of metal into the mold cavity. Advanced metal alloys and/or rapid solidification techniques can be employed to impart special characteristics to the casting. Ensuing heat treatment, straightening, and nondestructive inspection (NDI) techniques complete the process and yield a casting for final machining, surface finish, and assembly.

## IV. BELL HELICOPTER 427 PROGRAM

Bell's newest twin-turbine helicopter is much anticipated and is launched in a hungry market with high schedule compression. The program is perhaps the largest and most intensive CAD-based rapid-prototyping program ever driven in the aerospace industry. A total of 90 different new casting configurations were designed in aluminum, steel, and titanium alloys. A third of them were prototyped via stereolithography-based patterns in order to meet schedule requirements. Figures 5A–5D show the final products made via SL.

Due to time constraints and the size of the program, careful management and concurrent engineering were critical success factors. Quality, engineering and purchasing departments of both the vendor and the customer participated

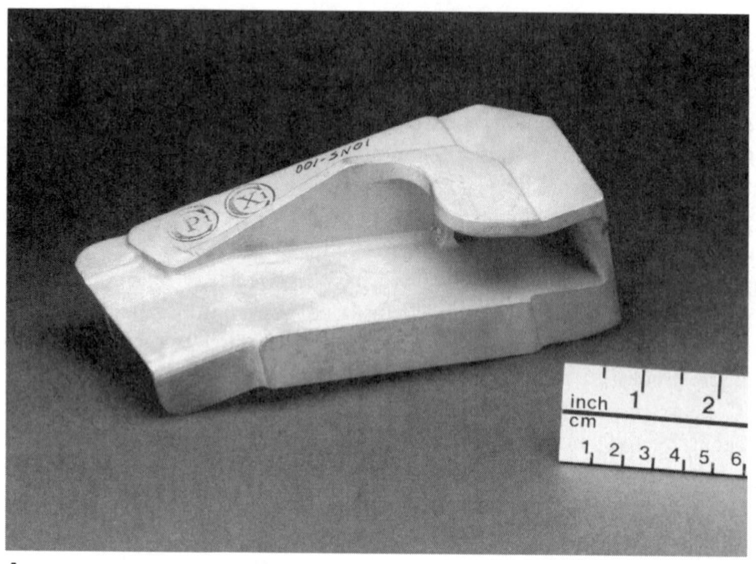


A

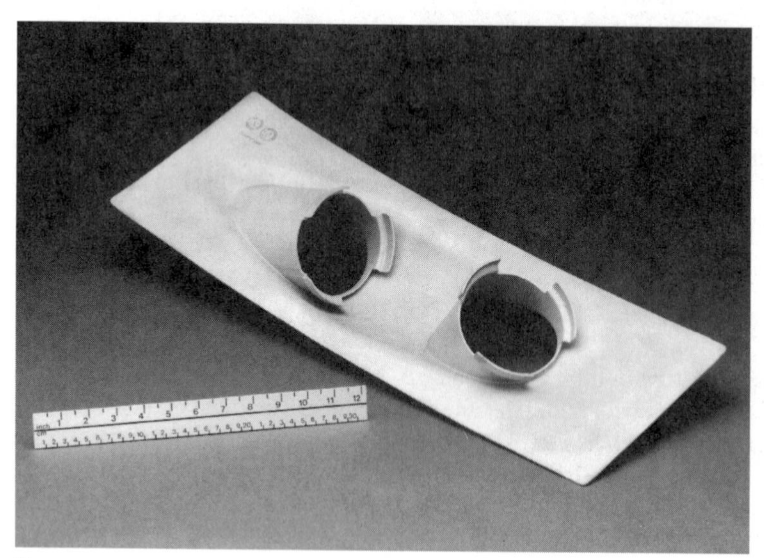

B

**Fig. 5** Final products made via SL.

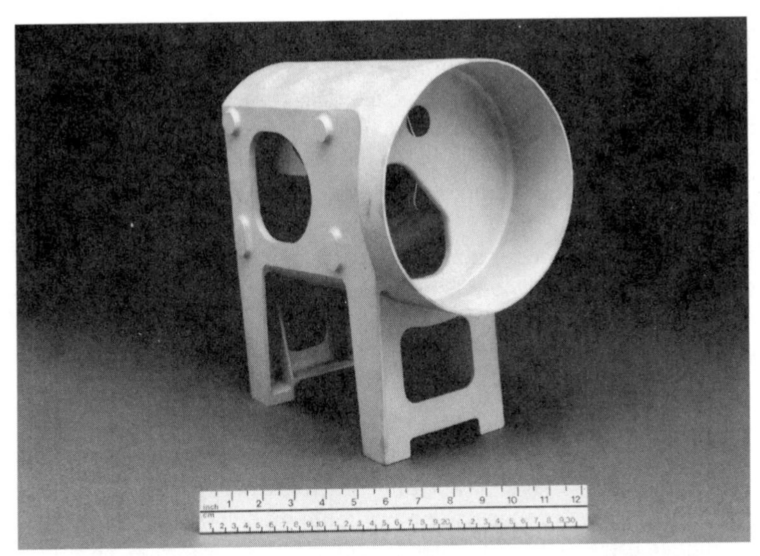


C

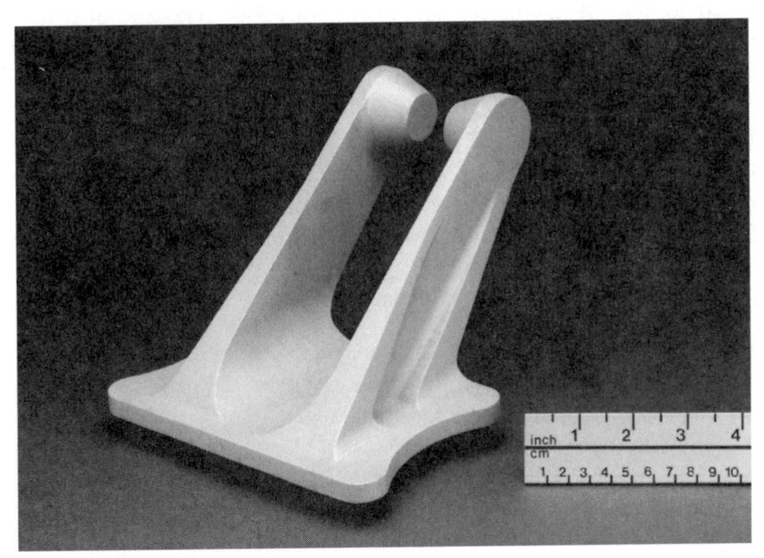


D

early and simultaneously. Quick reaction and responsiveness were achieved, in contrast to a normal response times of days-to-weeks for complex issues handled by large aerospace prime contractors. This proactive working relationship would identify challenges up front, minimizing the need to seek additional information, request for changes, and design changes once the program had entered manufacturing. Expectations for component finishing such as machining of close-tolerance features, surface treatment, and assembly are handled simultaneously. Today's aerospace prime contractors often seek a fully finished component rather than contract several sources, thereby eliminating additional orders and risking delays. Internet CAD file transfer has also cut numerous time-sensitive steps.

## A. Project Requirements

In the pioneering days of rapid prototyping, the foundry engineer was pleased to be able to demonstrate a successful transition from polymer model to metal casting without catastrophic mold-split, inclusion, or distortion problems. Yields were often less than 50%. Today, the technology has greatly matured, due in part to new RP model resins and pattern build styles and also with specialty designed gating systems, purpose designed shell mold processes, and unique mold firing techniques. Modern-day prototyping yields components with a >95% success rate, having predictable soundness, mechanical properties, and dimensional compliance. The mastering of pattern size variation and dimensional process optimization was the last of several key parameters to be accomplished. Requirements for modern-day aerospace prototypes are remarkably similar to certification of production hardware and may typically include the following:

- Chemical certification
- Casting microstructure evaluation
- Mechanical property certification throughout the part
- Radiographic compliance for soundness
- Pressure tightness in leak tests
- Surface finish validation and compliance
- Full 3D profile tolerance compliance

## B. Project Goals and Achievements

The main objective for Bell Helicopter Textron was to reduce the "time-to-market" and compress the procurement cycle for purchased metal structures.

Other important factors were also considered prior to investing in an RP center containing two SLA prototyping units at their facility in Fort Worth, Texas. Although ''machined-from-solid'' components may have yielded hardware for the first aircraft on some simple applications, investment castings were ultimately targeted for production aircraft, due to their great potential for cost reduction on the program. According to casting specialist Lloyd Lynch of Bell Helicopter, ''certification of medium size subcomponents can easily cost up to $50,000 (stress analysis, component overload test, fatigue test, vibration and heat/cold exposure trials). Bell's objective was to certify casting hardware on the first aircraft in order to prevent a re-certification as production castings later may have replaced the prototypes machined from solid or assembled components.''

The ability to quickly create prototypes which would be identical to hard-tooled production hardware later in the program was therefore a multiple bonus. In the rare case of failure of component fit or function changes, the RP manufacturing route offered Bell Helicopter a convenient vehicle for rapid change. The new design could then be retested within a few weeks and subsequently certified as required. Bell's requirements are indicative of the highest standards demanded from metal castings.

## C. Program Risk

Cercast has mastered the efficient transformation of a lightweight polymer RP pattern into a high-strength aluminum alloy casting with few technical risks. Aside from concurrently managing part design for producibility and determining capable tolerances for the component, one critical step remains. Gating technology and ceramic shell mold design, based on years of empirical design rules and experience, will improve the ability of the foundry to cast the component successfully with good soundness. Typical production techniques from traditional hard tooling often require several cycles of trial-and-error gate optimization to produce a defect-free and economical casting. However, prototype time constraints often require a usable component to be produced the ''first time around.'' A strategy of conservative gating and custom-designed shell mold system (which encourages directional solidification) has enabled the foundry to achieve a remarkable success rate, with few remakes being necessary on most designs. This is not to underscore those producibility discussions between partners on items such as tolerances, wall thickness, weld rework allowance, fixturing, and inspection aids that are all necessary for a successful program. Even with these requirements the fixtures and manufacturing aids will not be as complex and costly as those required for high-volume production runs.

*Solidification modeling* utilizes a numerical simulation of the casting and solidification process. Boundary conditions, shell characteristics, and other process parameters are used to simulate real metal filling and solidification conditions found in production investment-casting molds. This technique enables the foundry to determine the necessary locations to place gating attachments to eliminate feed shrinkage in the casting. Process modeling excels in demonstrating only the necessary attachment points, reducing the likelihood of excessive gating. Following the modeling, appropriate gates and runners are specified for production.

This analytical approach enables the foundry to develop and refine a gating strategy up-front, eliminating costly trial-and-error empirical testing. Accuracies of the model and computing speed are constantly improving, although the iterations can still be time-consuming and somewhat costly.

# 12
# The Future of Rapid Manufacturing

**Peter D. Hilton**
*Technology Strategies Group*
*Concord, Massachusetts*

There continues to be strong driving forces in industry to compete more effectively by reducing time and cost while assuring high-quality products and services. Some of these forces which will drive technology development and implementation in the area of rapid manufacturing are as follows:

1. Reducing the time and cost of new product development
2. Reducing the manufacturing cycle time
3. Reducing the cost of tooling to enable smaller economical lot sizes and, thus, product customization for niche markets or mass customization.

Several industries participate in annual cycles normally associated with seasonal sales around the Christmas holiday. We mentioned the toy industry earlier. The fashion watch industry is another example that is driving rapid tool development. Again, new generation products are needed annually. The faster the product development time, the later product development can be initiated and the closer to the market entry time the customer trends can be gathered and included into the watch design. Reduced development time is also very important for the automotive industry, which tries very hard to keep up with changing consumer priorities; for example, consumer preferences moved dramatically from small sporty cars to sports utility vehicles, leaving numerous automotive original equipment manufacturers (OEMs) scrambling to create products in this market niche (which is hardly a niche today).

Manufacturing cycle time relates directly to costs. By reducing the cycle time, one is able to produce more product with the same capital, as well as reducing labor costs per production unit. Injection molders compete directly on unit costs, and leading firms are very adapt at minimizing the cycle time (injection-molding machine time is often the largest component of the unit cost). They may use process simulation to assist in cycle time minimization (e.g., by performing design-of-experiment tests on the computer and thereby developing an analysis tool for process optimization). A major portion of the injection-molding cycle time is the time required to cool the part sufficiently so it can be removed from the mold without distortion. Approaches to enhancing mold cooling are included in efforts to reduce molding cycle time. One such approach is to incorporate conformal cooling channels into the mold, as discussed in Chapter 8. We predict substantial use of process simulation and conformal cooling to reduce injection-molding cycle time.

A major component of the cost of injection-molded parts is that for tool amortization. Obviously, the cost per unit goes up as the number of units to be produced in a tool decreases. This analysis has set minimum limits on the economical use of injection molding as well as other near-net-shape processes such as die casting. For smaller volumes, manufacturers have typically selected forming operations with lower tool costs and higher labor or other costs (e.g., machine capital). If tooling costs can be reduced, the equation shifts the minimum economic lot size for molding processes. This enables more customization for niche markets, shorter runs (and more product refreshment cycles), more product models, and so forth. Although reducing tooling costs is always of strong importance, the specific possibility of lower-cost tooling for shorter runs is technologically feasible. One is able to trade-off tool performance against cost (Table 1). Fortunately, it is likely that these lower-cost, lower-volume tools will also be able to be produced in less time. We anticipate an accelerating trend toward the development and use of lower-cost/shorter-life tools.

In 1993, we suggested a conceptual model as the target to strive toward. The model "Moldless Forming: An Advanced Manufacturing Process" was presented at an executive workshop with the same name, sponsored by Arthur D. Little, Inc. The idea was to envision designing products on a computer-aided design (CAD) system and producing them directly on some computer-controlled equipment without the use of any molds or special purpose fixtures. The team of industry leaders pondered the impact of such capabilities on their businesses. Today, we are getting a bit closer to achieving this paradigm, although we still have a long way to go. The concept is helpful for guiding the direction of research even while its full realization still eludes us.

**Table 1** Part Manufacturing Cost Elements

| | Tooling cost | Tooling development time | Tooling life | Cost per part |
|---|---|---|---|---|
| Traditional injection molding | $60,000 | 16–18 weeks | 250,000 parts | $0.24 |
| RapidTool injection molding | $20,000 | 6–7 weeks | 5,000 parts | $4.00 |

*Note:* Illustrative data from Anthony Anderson, Ford Motor Company.

An intermediate conceptual model "the disposable tool," is closer to reality. Imagine that the time and cost to produce tooling can be dramatically decreased. Then, one can consider use of the tooling to produce a lot of product and disposal of the tooling at the completion of the lot production. At another time, one could produce new tooling to produce more parts, and at that time, one might choose to update the product design at nominal cost. This approach would enable the user to avoid issues concerning different revisions of a product and concern about whether the tooling revision was consistent with the product revision to be produced.

We are very close to achieving this intermediate paradigm today. The direct use of stereolithography (SLA) produced mold cavity inserts in conjunction with standard mold frames has enabled the molding of severely limited (typically 5–50 parts) production runs. The run capability of the plastic molds is impacted by the material to be molded (filled and composite materials typically decrease mold life) and by the molding conditions (pressure and temperature). It is reasonable to predict continuing improvement in stereolithography materials as well as modification of other rapid-prototyping techniques to more closely achieve disposable tooling. On the other hand, some firms are working to reduce the severity of the molding conditions so that current SL mold inserts will be able to produce longer runs. A comparison among traditional tooling, disposable tooling, and moldless forming is seen in Table 2.

One encouraging development in this area is low-pressure metal injection molding. AlliedSignal has developed an aqueous binder called Agar which enables the formulation of feedstock that can be injected into a mold using a conventional injection-molding machine at pressures measured in hundreds of pounds per square inch rather than the usual pressures of several

**Table 2** Comparison Among Traditional Tooling, Disposable Tooling, and Moldless Forming

| Process | Steps | | | | | | |
|---|---|---|---|---|---|---|---|
| Traditional tools | Design part | Design tools | Make tools | Make parts | Store tools | Install tools | Make parts |
| Disposable tools | Design part | Design tools | Make tools | Make parts | | Make tools | Make parts |
| Moldless forming | Design part | Design process | | Make parts | | | Make parts |

*Note:* With traditional tooling, a design change requires a tooling modification which is costly and time-consuming. With disposable tooling, the tooling design change is made on the CAD system and the new tooling is made, as without a design change. The result is little cost or time impact associated with design change. With moldless forming, each part can be distinct at no additional manufacturing cost.

thousand pounds per square inch. The result is that soft tooling can be used for higher-volume runs and that ''disposable'' tooling produced by stereolithography can be used in mold frames to form hundreds to thousands of parts. These are metal powder parts and they require sintering after molding but the result is solid metal parts. This AlliedSignal technology enables the following paradigm shift: Metal parts can be produced using plastic tools as differentiated from conventional wisdom by which plastic parts are produced on metal tools. Of course, one can use this technology as a means to produce rapid (metal) tools.

More broadly, the desire on the part of product-development teams to have real prototypes (i.e., prototypes made from the production material by the production process) will drive continuing improvement of rapid tooling (or prototype tooling) technologies.

This desire is not frivolous; rather, it is based on the goal of easing the transition from design to manufacturing by verifying early in the product-development process that the parts can be produced by the anticipated manufacturing process. Further, this enables the development team to judge the tolerance capabilities of the fabrication process as well as to identify aspects of the design that may be difficult to produce. They can then make modifications to the product design or the processing to achieve robust manufacturing (the ability to produce parts within the required tolerances with a high degree of certainty). Robust manufacturing avoids high initial reject rates as well as early field problems. This conceptual approach to reducing quality problems is formalized through the use of statistics by setting allowable failure rates and designing the combination of the part and manufacturing process to assure that they are achieved. The terminology ''six sigma,'' which was first popularized by Motorola, refers to using the above approach to assure that the manufacturing processes include six standard deviations within the part tolerance band. The consequence of this method is that out-of-tolerance parts should occur at a rate of four per million parts produced.

As described in this book, there are several processes under current use as well as continuing development for making rapid tools that can be used to create ''real'' prototypes. These include direct stereolithography mold inserts, the use of SL or selective laser sintering (SLS) processes to form patterns for investment casting of metal mold inserts, 3D printing of ceramic shells also for investment casting of mold inserts, the various powder-based processes (e.g., Keltool), and those which involve deposition of a hard metallic layer over an rapid-prototyping (RP)-based pattern. Each of these processes is in limited commercial use today and development work is continuing on all of them. Although it is difficult to predict winners and losers among these technologies, we can confidently predict that rapid tooling will mature and that its use will spread over the next decade.

Georges Salloum wrote about process simulation in Chapter 2 of this book. Present analysis and simulation calculations have significant limitations that must be overcome in the future. Three-dimensional simulations of complex processes (e.g., injection molding or investment casting) may require hours to days on high-end computers. Further, the simulations while predicting trends and providing guidance are generally not sufficiently accurate to predict actual behavior. The inaccuracies are the result of both inaccurate input information on material behavior and approximations needed to reduce analysis times.

We believe that the development of computer software tools to support product development will continue and result in decreased need for paper or prototypes. Such CAD/CAE/CAM tools will continue to become more accurate and efficient as the power of desktop computers continues to increase. The result will be very fast responses for very complex calculations (e.g., simulation of the coupled fluid flow and heat transfer during the filling of a mold). The approach to product development will increasingly include CAD design, CAE analysis of performance, simulation (and optimization) of the manufacturing processes, and CAM, all using a single database and closely coupled. A bit further in the future, computing systems will be fast enough to enable real-time intelligent manufacturing process control (i.e., the process parameters will be monitored and compared to the optimal values as determined by the earlier analysis). The process will then be continuously adjusted to minimize the difference between actual conditions and optimal conditions. Alternatively, the processes may be managed by neural networks that enable learning and process improvement over time. Eventually, integrated computer-aided design, simulation, and control will enable combined optimization of product design and processing conditions, followed by actual processing at these conditions. The results should include product performance improvement, product-manufacturing cost reduction, low (or zero) manufacturing reject rate, and high product quality.

One specific area in which computer-aided process analysis will support process improvement is mold temperature control. One generally wants the mold cavity active surface to maintain nearly uniform temperature, independent of the particular process (injection molding, investment casting, etc.) so as to minimize part distortion and residual stress buildup during forming. Further, rapid transfer of heat from the part causes rapid part cooling and allows shorter processing cycles, saving capital and variable costs. Computer-based heat-transfer analyses can provide guidance on mold surface temperatures in terms of processing conditions and cooling systems. This information can guide the tool design for the location of cooling channels and the coolant flow rates to set for each channel. Advanced mold-making techniques such as that presented in Chapter 8 and named ExpressTool enable the construction of conformal cooling channels. The use of these advanced mold-making processes in conjunction with advanced analysis tools enables the creation of molds optimized to cause uniform part surface temperature during the processing cycle and rapid part cooling to reduce the cycle time.

Ideally, molds have active surfaces which are hard and abrasion resistant as well as able to withstand high temperatures and dramatic temperature cycles (just watch a die-casting operation in which molten metal and cold water sequentially contact the mold surface). On the other hand, the interior mold material should have high thermal conductivity to transfer the heat from the part and good fracture toughness to withstand the fatigue cycles to which it is subjected. This is traditionally accomplished through heat treatments and/or surface coatings. An advanced approach to achieving improved tools is to create ''gradient'' materials, that is, to somehow form a part with varying material composition (e.g., with a hard ceramic or cermet mold surface and a tough metal interior and a continuous transition between the ceramic or cermet and the metal composition). Gradient materials have been developed and formed by various deposition processes. Japan has been a leader in this area. The challenge that several rapid-prototyping technology developers are taking on is to produce gradient materials within the RP environment and, therefore, to enable the production of rapid tooling with gradient material compositions. For example, AlliedSignal is cooperating with Stratasys to form ''composite'' materials through the incorporation of multiple extrusion heads in their RP systems. By using loaded thermoplastics, this team is able to create a preform with various concentrations of ceramic and metal powders at various locations and can, subsequently, sinter the piece to form a mold insert with gradient material. Although these efforts are still under development at the time of this writing, they or similar ones are very likely to result in technology enabling rapid tool making with gradient materials. The Laser Engineered Net Shaping, or LENS process, currently under development at Sandia National Laboratories, Alberquerque, New Mexico, is also exploring the characteristics of gradient materials.

Although these various technology advances may occur at differing rates and having differing degrees of success, we can predict with a high level of certainty the overall trend to increased use of near-net-shape-forming processes and decreased use of machining. Net shape processes are more energy efficient and result in less material scrap. They can also be faster and less costly than the machining processes they substitute. Net shape process utilization is limited by the cost and fabrication time for the associated tooling. Reductions in both these factors will occur as a result of a combination of the technologies described in this book. The tooling will be further enhanced to contribute to process optimization through such factors as conformal cooling. The net shape (molding) processes themselves will also become more efficient through the use of computer-aided tools for process optimization, including process modeling (as discussed) and potentially neural-net or related techniques for continually learning and process fine-tuning.

# Index